STEALTH
AMATEUR RADIO

OPERATE FROM ANYWHERE

By Kirk A. Kleinschmidt, NTØZ

Published by
ARRL The national association for *Amateur Radio*
225 Main Street
Newington, CT 06111-1494

On Our Cover

Cover photo of Boston, Massachusetts skyline by Rob Schedinger of RJS Photography, Newington, Connecticut. Inset operator photo by Eric Swartz, WA6HHQ—Wayne Burdick, N6KR, enjoys making contacts with his Elecraft K2 from…anywhere! (Wayne and Eric designed the K2 as a full-featured HF radio that is fun to build and operate.) Inset highway and house photos from CorelDraw Clipart Photo Disk.

Copyright © 1999-2002 by

The American Radio Relay League, Inc

Copyright secured under the Pan-American Convention

International Copyright secured

This work is publication No. 247 of the Radio Amateur's Library, published by ARRL. All rights reserved. No part of this work may be reproduced in any form except by written permission of the publisher. All rights of translation are reserved.

Printed in USA

Quedan reservados todos los derechos

ISBN: 0-87259-757-1

First Edition
Third Printing, 2002

Contents

iv   **Foreword**

1-1   **An Introduction to Stealth Amateur Radio**

2-1   **Stealth Radio 101**

3-1   **Home is Where You Find Your Rig**

4-1   **Antenna Systems and Components**

5-1   **Stealth Antennas**

6-1   **Close Quarters: Handling Interference**

7-1   **Radio in Motion**

8-1   **Radio Destinations**

159   **About the ARRL**

161   **About the Author**

Foreword

The Rhombic Rangers want you!

I've always had a weakness for stories in which the forces of power and might get their comeuppance from those who are smarter and craftier. I cheer for the prisoners of war who build an escape tunnel under the commandant, and I root for the freedom fighters who slip in behind enemy lines. It's adventure of the best sort.

So I'd like to invite you on an adventure of your own—into the intriguing world of Stealth Amateur Radio. Kirk Kleinschmidt, NTØZ, a respected ham radio writer and former *QST* Assistant Managing Editor, is ready to guide you. A ham for well over two decades and a stealth operator for much of that time, Kirk has written a fascinating and useful volume that shows how to get the upper hand in the most trying circumstances. He's been there, done that, and has the QSL cards to prove it.

There may be forces of oppression surrounding you—antenna restriction laws, for example—but you can triumph through stealth and guile. Is there a dipole in that rain gutter? A vertical antenna hidden in that flagpole? A delta loop strung up in the trees? Only you know for sure.

If you believe that you need a few furlongs of aluminum sticking into the sky and a wad of cash to experience the best that amateur radio has to offer, I have a pleasant surprise for you: it's not so. There are many hams who have worked the world with shockingly modest stations. They'll tell you flat out that more bucks out of pocket do *not* equal more enjoyment.

What's more, Stealth Amateur Radio is *fun*. . .lots of fun! So it's no surprise that many hams voluntarily "go stealth" simply because they *prefer* it. It's enjoyable, challenging, and low cost. What's not to like?

If you've been looking for more fun, a challenge, or just to get on the air inexpensively, strike a blow for freedom: hang a rhombic or loop in the trees, dangle a longwire by the downspout, clamp a whip to the balcony rail. A world of adventure awaits you.

Jock Elliott, KB2GOM

Chapter 1

An Introduction to Stealth Amateur Radio

If you're like most hams, you've spent at least a little time daydreaming about having the "ultimate station"—what it would do for you, how it would make you feel, the ease with which you would work even the weakest stations and so on. A stack of monobanders at 100 feet, inverted V dipoles for the low bands, three transceivers, a legal-limit amplifier and an espresso machine, perhaps?

Well, not exactly.

Although you might even make section champ in a contest or two with such a setup, you can find stations of this caliber just about anywhere.

Because we're already daydreaming—and so early in the book!—consider a more robust fantasy station. For example, the particular installation I'm imagining has an antenna farm so vast and so powerful that virtually no signal escapes its attention—or its wrath. Inside, the walls are paved with rack-mounted amplifiers, and the glaring smear of light from the panel meters—six per amp—illuminates the operator's position like a night game at Candlestick Park. The power cable that feeds the Amplifier Monolith is a helix of three massive strands, each as thick as a wrestler's forearm. A 60-Hz ac hum fills the room with subconscious anxiety…

You get the idea. Your dream station is probably a bit different, but the net result is the same. While operating these comic book "superstations," previously impossible pileups part like the Red Sea, CQ DX calls result in immediate pileups regardless of band or time of day, other hams complain about bent S-meter needles, and propagation rules that bind others have little meaning to you. In short: One *heck* of a station.

The Dream Stations we fantasize about represent Amateur Radio's version of the Great American Dream. (Hams in other countries can substitute as necessary!) Some farmers have garden plots, and some have spreads the size of Rhode Island. Some boaters scoot around the bay in nimble runabouts, while others pilot triple-hulled oil tankers. Accordingly, some hams—most, probably—have modest stations, while a few enjoy honest-to-goodness Dream Stations. Luckily for hams everywhere, we can all *enjoy* ham radio regardless of our means.

Although my own stations have always been in the Stealth or Stealth-plus categories (unless I happened to be operating from

---

RF EXPOSURE

Amateur Radio is basically a safe activity. In recent years, however, there has been considerable discussion and concern about the possible hazards of electromagnetic radiation (EMR), including both RF energy and power-frequency (50 - 60 Hz) electromagnetic fields. FCC regulations set limits on the maximum permissible exposure (MPE) allowed from the operation of radio transmitters, including Amateur stations. These regulations do not take the place of RF safety practices, however.

Most amateur stations easily comply with the rules, but indoor antennas, antennas mounted close to houses or people or high power operation can sometimes result in exposure in excess of the permitted limits. The ARRL has published a book on RF exposure, *RF Exposure and You*, which covers the subject in detail. You can learn more about what is in this book or order it online at: **http://www.arrl.org/catalog/6621/**. The ARRL also has a Web page that contains an overview of RF exposure, with links to many valuable sites: **http://www.arrl.org/news/rfsafety/**. Amateurs should read and understand this information before using high power or antennas that are located close to people. — *Ed Hare, W1RFI, ARRL Laboratory Supervisor*

W1AW), I came up through the ham radio ranks as a teenager, often drooling over and daydreaming about the awe-inspiring stations glorified in *QST*.

Like ham radio John Waynes perched on their trusty steeds, the Dream Station owners shown in the ham magazines in the late '70s were usually Old-Timers who posed like elder statesmen in the midst of their all-encompassing shacks. They'd had decades to string 14 rhombic antennas between pitch-tarred pine poles so tall they're now extinct, buy oceanfront land for $50 an acre and assemble a wall full of rack-mounted Collins radios. Like all larger-than-life figures, these operators lived in the rarified air at the top of the DXCC Honor Roll and were universally respected by men, women and the heads of foreign governments. The stations they called *always* came back and their DXpedition exploits filled many a magazine.

Who could remain immune to the power of these images? I certainly couldn't!

In high school, while killing time in boring classes, I would often doodle around with "designs" for ridiculous antennas. Perhaps I'd be the first ham on the block to have a full-size, 6-element, 160-meter Yagi at 300 feet? The tower and antenna would both be made from a boron-reinforced carbon-fiber skeleton that I could fabricate for no more than $56—my pay for two weeks of cleaning up the local bakery after school.

When I received a program guide from Radio Netherlands, I threw *my* radical designs out the window. Obviously, I wasn't thinking big enough. In Hilversum, Holland, as the color brochure revealed, the antenna farm of my favorite shortwave broadcaster sported several rotatable log-periodic beams mounted on huge monolithic spires. As I recall, these monsters covered 2 to 30 MHz (continuous) with 18 full-size elements on a 150-foot-long boom that was big enough to *walk inside*. I could imagine nothing better, and if I had one I'd gladly put it in my backyard!

In drafting class, all of my poorly rendered three-bedroom ramblers were nestled into picturesque cul-de-sacs and came factory equipped with big, beautiful antenna towers topped with stacks of monobanders. I paid special attention to getting the perspective of the Yagi elements just right. I thought (and still think) that ham antennas were beautiful (eggbeaters and helical types excepted). Their symmetry and progressive scaling have a simple, mathematical elegance.

Phased verticals U-bolted to the chain link fence? Beautiful. A Sterba curtain strung between towering pines? Lovely. Slopers skewed all around the backyard? Perfect. A menagerie of mix-n-match coax cables snaking into the foundation crack by the basement window? As pretty as the twisted roots of a bonsai tree.

Not only were these things beautiful in and of themselves; as a ham, I knew that they had a functional beauty as well. In an era when the Internet was used only by scientists, and long-distance telephone calls were expensive and uncommon, ham antennas allowed domestic and global communication on a personal level. They chatted with friends and brought exotic, faraway hams to my basement shack. They produced stacks of foreign QSL cards, many filled with interesting foreign scripts.

Antennas *made* ham radio happen. They were the *key* component. If you had a fabulous antenna and a garden variety rig, you were still in the game! The same is true, today, of course, but if you're reading this book, it's reasonable to assume that you

This big, beautiful log-periodic antenna belongs to Radio Sweden. Although I would love to have it in my backyard, I'm sure my neighbors wouldn't be as enthusiastic! (Photo courtesy Radio Sweden and Swedish Teracom)

know that many people think ham radio antennas are nothing more than eyesores and property "devaluators."

ENCROACHING REALITY

This is an ironic time to be a ham. There have never been more ham operators. Radios have never performed better or been more affordable. Computer technology makes operating, logging and contesting a snap. Digital signal processing makes impossible signals copyable, and the Internet provides a constant stream of information and Amateur Radio resources. As I'm writing this, the sunspot cycle is cooperating and the HF bands are awash with easy DX pickings.

On the other hand, from a municipal perspective, setting up and using even a modest ham radio station is no longer a no-brainer. There are fewer open spaces, and populations—especially in metropolitan areas—are steadily increasing. Despite the desires of average citizens, government seems to be looming larger in our everyday lives. Lawsuits are a new national pastime. Stress is increasing. Laws, taxes, ordinances, rules and regulations abound. We even work more and play less!

This *encroaching reality* plays havoc with ham radio, too. Your neighbors—whom you probably don't even know, unlike in decades past—aren't likely to care one whit about your radio hobby. The fact that you're a federally licensed *emergency communications resource* doesn't even register (unless there's a flood or hurricane, of course). The adage about a ham's home being his/her castle was easier to realize in the Golden Age of Radio—before society ran amok.

Your antennas *do* register, though—probably in the negative. Your annoying transmissions, whether you make them or not, allegedly cause hair loss, attract ball lightning and implant subversive messages into the minds of youngsters. They're also credited with messing up the TV, the telephone and the VCR—which, in fact, they might!

An average neighborhood committee sees no beauty in the symmetry of your log-periodic array. Most neighbors don't care that there are talkative hams on Pitcairn Island—and they couldn't find the history-laden dot on the map if they had to. In their eyes you're just a freak wave in a sea of interference-prone consumer electronic devices.

CC&RS

To keep radio hobbyists at bay, those responsible for housing developments, subdivisions, annexations, apartment complexes, neighborhoods and even entire communities have enacted prohibitive covenants, conditions and deed restrictions (CC&Rs)—not to mention overachieving city and county ordinances. When purchasing your house, apartment or condo, if you sign on the dotted line—aware or otherwise—your external antennas may wind up being a thing of the past (the non-Stealthy ones, in any case).

To help Amateur Radio operators effectively deal with restrictive local (non-federal) ordinances, in 1985 the FCC enacted something called Memorandum and Order in PRB-1, "limited preemption of local zoning ordinances." PRB-1 set rules for local municipalities to follow when regulating Amateur Radio-related antenna structures. To further assist hams, the essence of PRB-1 was incorporated into the FCC's Part 97 Rules in 1989, effectively strengthening the FCC's sway over municipalities.

Although PRB-1 has been a big help to hams, it's useful only when dealing with local government ordinances—and property affected by CC&Rs isn't "protected." Covenants, deed restrictions and bizarre ordinances can be circumvented in some cases, but the specific remedies are beyond the scope of this book.

To learn more about PRB-1, visit the ARRL's Web site at **http://www.arrl.org** and go to the Regulatory Information section or contact the Regulatory Information Branch at ARRL Headquarters and request the "PRB-1 package." (There is a charge for this package.) Additional material and the complete text of PRB-1 can be found in *The ARRL's FCC Rule Book*.

STEALTH RADIO REVEALED

This book is about the broadly defined topic of Stealth Amateur Radio—exactly how to get on the air and enjoy ham radio, whatever your situation and whatever your motivations.

The information is useful to hams in almost any situation—Stealthy or otherwise—and its practical solutions encompass a wide variety of "belief systems." Although I use the term Stealth Amateur Radio (or Stealth Radio), you may prefer to think of what we're doing here as Low Profile Amateur Radio, Low Impact Amateur Radio, Politically Correct Ham Radio, Can't We All Just Get Along Radio, Good Neighbor Radio, Real Estate Friendly Radio, Spouse Accommodation Radio, Picturesque View Preservation Radio, Traveling Salesperson Radio, Spending Winter in Arizona Radio, etc. Fill in the blanks appropriate for your situation.

ARRL founder Hiram Percy Maxim, 1AW, didn't have to worry about zoning ordinances when he put up this impressively large antenna at his Hartford, Connecticut, home in the 1920s.

Stealth Radio, like ham radio in general, encompasses a range of station setups, from barebones to extravagant. Some Stealth operators feed $7 attic antennas with a pair of $3000 transceivers. Some connect an invisible outdoor antenna to a $10 home-brew station. Most use something that's in-between. *Any* of these stations is legitimate "Stealth Radio."

PHILOSOPHY 101

The philosophy behind choosing a Stealthy approach to ham radio is important—and will likely become even more so as Amateur Radio moves into the uncharted territory of the 21st century. You might be forced by circumstance to adopt Stealth techniques, but other hams are *choosing* a low-impact approach to hobby radio. These choices parallel conservative and "harmonious" trends in other areas of life.

In these hectic times, many people are choosing to simplify, simplify, simplify. Bookstores are filled with paperbacks that show us how to get off the fast track—the same fast track everyone was struggling to get on little more than a decade ago. Back to the land, with upwardly mobile twists, perhaps, is a *bona fide* trend. Retiring early to play golf or go trout fishing is no longer a rarity. The list goes on—and ham radio isn't excluded.

You have to admit, too, that low impact radio stuff is easy to take care of. It usually costs less. It doesn't topple over in ice storms. It doesn't block the neighbors' view of the mountains. You don't have to give it much thought. If someone steals it, it's no big deal. It doesn't spin the electric meter faster than a speeding clothes dryer, either.

AMATEUR RADIO'S MIDDLE PATH

Why would otherwise normal hams choose to set up minimalist, low impact stations with Stealth antennas (indoors, perhaps) and low power levels? There are many reasons. Here are only a few:

THE AMATEUR RADIO SERVICE

Amateur Radio operation in the US is constituted as a radio *Service*, with rules, regulations and goals that go beyond the interest of mere hobby operation. In becoming licensed hams, we agreed to play by those rules. One of the most important Service rules compels us to use the minimum transmitter power required to communicate.

That doesn't rule out the use of large, powerful stations, of course, but it does offer the opportunity to embrace low impact radio—as many hams do. Running high power to a giant antenna array doesn't necessarily violate the rules, but it's certainly overkill when it comes to chatting with the gang across town (or when propagation clearly doesn't require it).

The minimum necessary power rule and the doctrine of *low impact* operation protect us all. They promote responsible, considerate operation. Try it sometime! Reduce your 100 W signal to 50 or 25 W. Thanks to years of low-power Stealth operating, I *know* that you'll maintain effective communications *most* of the time. You'll also improve your operating skills, enjoy a greater sense of achievement and gain an intuitive sense of propagation.

SKILL VERSUS BRUTE FORCE

Long before David and Goliath had their epic battle, skill has been tangling with brute force. I'm sure you have your favorite analogy. Basically, it comes down to the fact that any idiot can fire up a water-cooled Voice of America-size transmitter and blurt out a whopping signal. I place hams who take this approach by default in the same category as the guys who screech the tires on their pickups or water their lawns during drought emergencies. Both are equally *consumptive*.

On the other hand, those who choose the low profile lifestyle align themselves with the Davids of the world, substituting skill and persistence for brute force. They're in good company—and they're upholding the tenets of the Amateur Radio Service in their own special way.

THE GOLDEN RULE

Hams treading the Middle Path are concerned about others—hams, neighbors and family members. They try to fit in, to get along, to accommodate a community of interests in addition to their own. They practice the Golden Rule: Do unto others as you would have them do unto you (reasonable variations notwithstanding).

As hams who comprise a federally licensed emergency service, we enjoy certain protections from *unreasonable* local restriction. These privileges are welcome and necessary as a whole, but they can be easily abused.

Just because we *can* transmit a 1500-W signal doesn't mean we *should*. Just because we *can* erect a 200-foot-high antenna tower doesn't mean we *should*. Hams who follow the Golden Rule *integrate* their radio pursuits with the pursuits of others—not because they *have* to, but because they *want* to! Governments can't legislate common sense. That's up to us.

WHO ARE STEALTH OPERATORS?

Hams who embrace Stealth Amateur Radio do so because of a wide variety of circumstances and philosophies. Do you fit into any of the following categories?

- Apartment/condo dwellers and those who live in neighborhoods that are affected by covenants, conditions and restrictions (CC&Rs).
- College and tech school students who live in campus dormitories, apartments or condos.
- Travelers, truck drivers, salespeople, campers, bicycle tourists, snowbirds and RV jockeys. Ham radio is the perfect companion for "chronic travelers."
- Those who want to make ham radio a *part* of their lives instead of an overwhelming pursuit.
- Hams who don't want to tangle with their neighbors—even if the hams know they're operating in strict accordance with FCC regulations and are clearly "in the right."
- Hams who'd rather not see *their own* antennas!
- Veteran hams who are "burnt out" and need a fresh approach to the hobby.
- You, perhaps?

IT'S ABOUT ENJOYING RADIO

Regardless of your "Stealth circumstances," it's safe to assume that you want to enjoy Amateur Radio fully and make the most of your situation. With the information in this book you can do just that.

Uncounted thousands of hams worldwide rely on Stealth

techniques to get on the air. Although they may not mention it during QSOs or print it on their QSL cards, if you dig a little deeper you'll be surprised at the number of Stealthy contacts you've probably already made. The fact that the other operators were being Stealthy—and you didn't even notice—illustrates my point perfectly!

Unless you're an experienced Stealth Operator, I'd suggest reading the book from cover to cover. By the time we're through you'll learn:

- How to set up and operate a Stealth Amateur Radio station from home (house, condo, apartment, whatever).
- How to successfully operate with low power (maybe even QRP!) and less-than-optimum antennas.
- What's required to make sure your indoor ham radio installations are safe.
- How to install a wide variety of Stealth antennas—at least one for every situation!
- The basics of mobile HF and VHF antennas and operation.
- How to have fun operating while portable. You can use these skills and techniques for camping, vacation DXpeditions, VHF/UHF mountaintopping, Field Day and more.
- What to do about interference.
- About a variety of "Stealth Alternatives"—ham radio activities you can enjoy without a conventional station of any kind. Enjoy these individually or use them to enhance your traditional radio pursuits.
- Where to find additional resources, support and information.
- How to get on the air from just about anywhere or any place and have fun doing it!

Stealth Amateur Radio is about having fun, adapting to unique environments, overcoming obstacles, learning about new and interesting facets of Amateur Radio and understanding that you *don't* need to conform to decades-old stereotypes to have an awful lot of fun.

Be persistent. Be creative. Keep an open mind. Let's get Stealthy!

Chapter 2

Stealth Radio 101

Before starting your Stealth Radio adventure, take the time to really understand your situation: operating conditions, physical situations, resources—everything that affects your station and the kind of operating you can or might want to do. Before deciding on the specifics, try to explore a variety of options to come up with a personalized Stealth Solution.

It'll probably help to make a list of the ham activities that you find the most interesting, and any obvious physical considerations. For example: Is home operation necessary, or would a mobile rig in the car be better? Are you a homebody or a long-haul truck driver? Is a local club station available, or are foxhunting on weekends and public service communications really what the doctor ordered? Does your radio buddy have a killer contest station a few miles out of town? How about putting together a compact station for the RV?

If home operation is mandatory, investigate antenna possibilities and possible shack locations. Is there any way to erect a modest non-Stealth antenna? Would a chat with the landlord possibly help? If you did put up a modest "visible" antenna, who might complain? What consequences might result? Do you even have room for a compact ham station in the spare bedroom, the closet or the laundry room?

How much money are you willing to spend? Does your budget call for a used transceiver, a few accessories and an inexpensive hidden antenna? Or can you afford a deluxe "off-site" station that's remotely controlled from your house or car?

Frequency also plays an important part in determining Stealth specifics. Is HF operation necessary or do you simply need to hit the local repeaters? With a few exceptions (notably moonbounce and a few other weak-signal activities), setting up Stealthy VHF and UHF stations is fairly straightforward. Building an effective Stealth antenna for 160-meter work, however, can be quite a challenge.

Don't forget about philosophy! You might choose Stealth Radio to preserve your own sense of aesthetics!

STEALTH ALTERNATIVES

While you're sorting through your Stealth Radio options, be sure to consider activities and alternatives that might otherwise go unexplored. Some of the activities in this section are mainstream Stealth Radio pursuits (and are covered in more detail elsewhere in the book), while others can be enjoyed in addition to (or at the same time as) a more conventional approach to the hobby. Any could turn out to be your favorite radio specialty, so don't write them off until you investigate further!

CLUB STATIONS

Unless you live way out in the boonies, there's probably a club station (or two) in your area. Amazingly, it's probably underutilized, waiting for you to lovingly twist the knobs and twirl the antenna. Although popular in Europe, where some countries require a period of club station operation as a licensing requirement (or as an option for those who hold lower-class tickets), club stations in the US are often used primarily for license instruction and contesting.

One club station near my town has four transceivers (two HF, two VHF), a large amplifier, a tall tower, rotateable beams and

The club station at the Pavek Museum of Broadcasting is a bona fide superstation—1968 style. The fully functional station serves as a meeting place for many Twin Cities' radio clubs and electronics societies. (author photo)

decent wire antennas for the low bands. The kicker is, it's hardly used! If I wanted to get on the air while I was sorting out my own Stealth Situation, that's where I'd start. I'd have the place almost to myself!

Search for club stations in the usual places—ham radio clubs—and at colleges, universities, tech schools and even high schools. You might have to join a club or volunteer your time helping out at a school station, but that's just another part of Amateur Radio.

RADIO IN MOTION

Even if you have the most restrictive CC&Rs imaginable at home, chances are still very good that you can move your shack into the car (boat, RV, motorcycle, etc). Mobile hamming has come a long way since the '50s and '60s, when mobile rigs were the size of small refrigerators, required power supplies that were even bigger, and used mobile antennas that made your car look like a salvage yard.

Mobile HF operating is enjoying a real renaissance. Today's mobile radios are small, small, small! They're also full-featured,

Former ARRL HQ staffer Al Brogdon, now W1AB, has two main hobbies: Ham radio and touring motorcycles. Thanks to a little ingenuity and a bit of modern technology, he can enjoy them simultaneously!

perform well and don't cost an exhorbitant price. Because they run on 12 V dc, you can use them as the centerpiece of your indoor Stealth station, take them camping, use them for Field Day, fly them to the Caribbean, etc.

Many modern mobile HF radios handle AM, FM, SSB, CW and data modes; cover 160 through 6 meters (or more); and receive from dc to daylight. Some can even be remotely mounted (the radio lives in the trunk or under the seat while the "control head" and the mic mount to the dashboard). These little rigs are nothing, if not flexible! They provide many Stealth operators with "one rig for all occasions."

Mobile antennas, once shrouded in myth and mystery, are now known quantities. They're still not terribly efficient when compared to larger antennas (especially on the low bands), but at least we know why—and how to make them perform to the best of their abilities! The basics of mobile hamming are covered later in the book.

CONTESTS, FIELD DAY AND SPECIAL EVENTS

When I first moved to Connecticut I was surprised to find a veteran ham who didn't own a station, yet operated almost every weekend. During the following workweek we'd hear about his contest exploits. This fellow worked the world from his friend's "contest superstation," which was lavishly equipped and advantageously located in the low mountains of western Massachusetts. When I asked him why he didn't have a station at home, he merely gave me a curious glance. This was exactly his kind of ham radio. He wasn't missing out on anything.

This weekend warrior wasn't alone, either. I soon met others who took similar approaches. Some operated from friends' stations, some from W1AW, some from university club stations. One guy got on the air only when he was vacationing on one tropical island or another!

If this style of ham radio appeals to you—and you have generous friends or an available club station—you'll have plenty of operating opportunities, including Field Day and Special Events stations.

SHORTWAVE LISTENING AND SCANNING

These ham-related hobbies are often credited with bringing people to Amateur Radio, but there's no reason why we can't participate in all of them simultaneously. I've never met a shortwave receiver or scanner that could mess up the neighbor's TV (certain regenerative receivers excluded). Thanks to the wide-range receivers found on most modern ham rigs, Stealth SWLs won't even need an extra radio.

SWLs and scanner enthusiasts listen in on local public service communications, communications from airplanes or ships at sea, a huge variety of international broadcasters (familiar and obscure), clandestine and pirate stations, and much more. Fun stuff!

FOXHUNTING AND RADIOSPORTING

Foxhunting—finding hidden transmitters as part of a friendly competition—is a popular weekend activity in many parts of the country (especially on both coasts and in larger metropolitan

David Pingree, N1NAS, tests a foxhunting rig in the back yard at ARRL HQ.

areas). Hams, usually radio club members and often grouped in age- or experience-related teams, gather to search for one or more hidden transmitters (foxes). The search area may be as small as a schoolyard or as big as a state!

On a typical foxhunt, competitors try to find all of the foxes in the least amount of time. Common frequencies are on 2 and 80 meters. Competitors use hand-held radios and compact directional antennas. Larger competitions may cover several square miles of forest or park land and may require maps and orienteering skills.

In the "motorsport" variant, the hunters drive cars or off-road vehicles, the foxes are typically hidden on mountaintops or wayside rest areas, and the field of competition may cover several hundred square miles. Mobile foxhunters often use GPS navigation systems and sophisticated receiving gear, including multi-antenna Doppler arrays with computerized graphical displays.

Whether the atmosphere is casual or highly competitive, foxhunting has something for everyone! Contact the clubs in your area to find out who sponsors foxhunts, then show up and ask to join one of the regular entrants to see what goes on.

PUBLIC SERVICE COMMUNICATIONS

Providing communications at public events—parades, celebrations, and so on—is a long-held Amateur Radio tradition. Although FCC rules prohibit amateurs from relaying certain specific information about race leaders and other information on the progress of an event for the benefit of event organizers, hams may assist safety officials at aid stations, operations centers, checkpoints and emergency vehicles.

To get involved, all you need is a hand-held transceiver. Most public service communications are handled on VHF and UHF frequencies because few activities spread out beyond simplex or repeater range. Two meters is most popular, but other bands are also used.

If you're a member of a ham radio club, you've probably already been asked to help out at public events. If you aren't in a club yet, or if your club hasn't engaged in such activities, ask around on the air and check the local nets to hook up with service-minded hams in your area.

In addition to local club-provided communications, hams must be prepared to handle larger regional or national emergencies such as floods, fires and earthquakes. Most of these emergencies are handled by members of the Amateur Radio Emergency Service (ARES) and the Radio Amateur Civil Emergency Service (RACES). Other popular public service outlets include SKYWARN. Its local chapters spot and track tornadoes and other severe weather conditions, and often work closely with the National Weather Service.

If you want to serve your fellow citizens, public-service communications will provide the opportunity—no home station required!

EXPEDITIONS

You can't be the first explorer to reach the North Pole, but you can take your radio gear to an infinite number of enjoyable "expedition destinations" that will definitely be appreciated by your fellow hams.

Where might you go? Just about anywhere. How about camping, canoeing or motorcycling? Or maybe fishing, hunting or hiking. Don't forget Field Day! With a compact mobile rig or

Not all expeditions end up in remote, exotic destinations. For VHF/UHF contesters, a nearby mountaintop is the place to be.

an even-smaller QRP transceiver, you can be on the air from just about anywhere. Stay in touch with friends and family, make new friends, or both!

These activities might be fun, you may be thinking, "But why will they be appreciated by other hams?" Well, if you add a few extra elements, it will be clear.

During Field Day or the November Sweepstakes, for example, instead of operating from New Jersey, which has scads of hams, why not take your camper—and your radio gear—down the road a ways to Delaware, where hams are scarce and sought-after? By working the contest from a rare state you'll be "the DX station," and others will be appreciative! Every year at least a few Alaska hams trek across the border to work Sweepstakes from Canada's Yukon or Northwest Territories. Why? To *be* the DX, of course!

Your expedition activities don't have to be limited to contests:

• Setting up at a scenic overlook at an out-of-the-way mountain

pass can help other operators collect a new grid square.
- Operate from a backwater county and other county hunters will think you're swell (county hunters often use specific nets to coordinate their activities).
- If you can, operate from a nearby island (inland or coastal). Operators looking for Islands On The Air (IOTA) QSOs will be looking for *you*.
- Set up at any state capital and folks on the 3905 Century Club net will want to put you in their logs.
- When vacationing at any "DX" locale, pack your mini mobile rig, a wire antenna and a foreign license or reciprocal operating permit and you'll see what it's like to be on the other end of a pileup!

INTERNET "HAMMING"

Purists may swoon over this one, but there's no denying the value and utility of the Internet when it comes to being a ham. In addition to the flood of radio information, expertise, how-to articles and manufacturer contacts, the web offers its own ham-like activities: Internet chats.

You can chat mic-to-mic, keyboard-to-keyboard—even camera-to-camera—with hams and nonhams alike, domestically or overseas, 24 hours a day—the Internet has perfect propagation. I ignored this avenue for years until I heard a friend excitedly babbling about talking to a student in Israel and a car mechanic in Italy on the Internet. "It was just like talking on the phone," she said, "just like being there!" Just like ham radio...

When you get right down to it, talking is talking.

REMOTE STATIONS

Once pursued by the technically elite, remote stations are now off-the-shelf items. They're not inexpensive, but they can provide a superior Stealth Radio solution for some operators. Instead of hassling with trying to put up a station at your new deed-restricted home, why not put the station—and a full size antenna or six—somewhere else?

This has been done before, of course, but the control links were complicated and the flexibility and facility offered by the radio and control systems were usually limited.

Stacked on top of this Kachina 505DSP computer-controlled Amateur Radio transceiver are a 12-V power supply, a remote-control box and an automatic rotator controller.

As I'm writing this in 1999, Kachina's DSP505 transceiver stands out as a turnkey solution for operators interested in full-function remote stations. In fact, the '505 is designed expressly for remote operation. It's basically a computer-controlled black box with no knobs or readouts of any kind.

Base station users set their PCs alongside the '505 and operate the rig using the computer and a monitor to control the radio. Remote users have a PC and a Kachina remote-control link (plus a mic and keyer paddle) in their shacks. They "call" their remote station on the telephone to get on the air! No Stealth is required. There's also no RFI and no lawsuits filed on behalf of the neighborhood association. The transceiver and a second remote-control link are set up at the remote site.

This allows some interesting scenarios. If you had such a system, you'd likely build your remote station fairly close to home—at least in the local calling area to avoid paying long-distance charges every time you "call your radio." If you wanted to place your remote station halfway across the country—or even in another country, assuming reciprocity and third-party agreements could be worked out—the only limiting factor is the cost of the telephone calls while you're on the air! Even that hurdle may have solutions, though. How about an Internet link between you and your remote station?

If you live in New York, say, and you have a Kachina Buddy

in Los Angeles, you can fire up *his* station from *your* shack as long as he's provided the appropriate security codes. Clubs and contesters are just starting to take advantage of this technology.

With a system cost of about $2500 (again, in 1999), many Stealth operators will be priced out of the market. But when you consider that hams routinely pay $10,000 or more battling ordinances and CC&Rs, the remote systems may be a bargain. Prices will likely fall as the technology matures.

Other remote systems let you control and operate a remote radio with a conventional VHF ham transceiver and a suitable control interface. These systems tend to be less expensive, but they may require specific radios and may not function on all modes, etc.

THE STEALTH POWER CONSTANT: HOW LOW CAN YOU GO?

For many reasons, most Stealth Radio operators will be limited to low power (by circumstance or by choice). That's why I call it the Stealth Power Constant. It's as universal as Avogadro's Number or Fibonacci's Series (or would you prefer as sure as death and taxes?). Unless you're a glutton for punishment, forget about feeding invisible antennas—especially indoor antennas—with legal-limit amplifiers.

What happens if you do? Lights dim, TVs blink and bobble, microwave ovens beep disconcertingly, computer power supplies fry, the neighbors become hostile, the VCR stops recording—and you'll likely expose yourself and your loved ones (well, loved ones and *neighbors*, anyway!) to potentially dangerous levels of RF energy.

My "power folly" was a home-brew kilowatt amplifier feeding a horizontal loop antenna that encircled the house just under the second-floor eaves (the stand-off insulators were eight inches long). Because my shack was also on the second floor, my feed line was only three feet long. Although I didn't have to worry about high SWR levels on my stubby length of coax—and I achieved my goal of working Australia on 80-meters—I also worked everything in the preceding paragraph, and more. After all, the whole house was "in the antenna." (That was before all the studies about RF energy heating body tissue and long before the EPA and the FCC established maximum exposure limits. I would definitely not repeat that experiment today! See the **RF Exposure** sidebar in Chapter 1.)

By the way, the Horizontal House Loop worked very well at friendlier, lower-power levels, QRP included. DX on the high bands was a sure thing, and after I disassembled my twin 4-400A amplifier (which was collapsing my operating desk), 80-meter DX QSOs still made it into the logbook. I can vividly remember working UAØs on 80 CW. I was running 80 W to a 40-meter loop that was 15 feet above the ground and tacked under the eaves of a bright green house that I rented with several college buddies.

Encouraged by the low loop's performance, I put up a super-Stealth version supported by two trees in my yard, and one tree *across the street* in a neighbor's yard! This horizontal loop was up pretty high, 50 feet at least. It was made from No. 24 *steel* wire (my choice for durable invisible antennas at the time), fed by 50-Ω coax and supported by lengths of clear Weed Eater line (absolutely invisible once airborne).

Although I probably violated several ordinances, and antennas that span public roadways (but not power lines!) certainly don't conform to the FCC's definition of good engineering practices, I used the antenna for nearly a year before taking it down. I never did reassemble the amplifier. It just wasn't necessary.

Unless you can hide your Multiband Stealth Dipole 40 feet off the ground in a swampy, wooded area a hundred feet behind your house, like one semi-Stealth operator I recently encountered ("Hey, I can run a kay-double-you to my 'hidden' antenna and not bother anyone" [Duh!]), your power output should stay between 10 and 100 W—but feel free to run 1 W or 5 W if it's sufficient. Yep, that's right!

To minimize interference and RF exposure, get used to the idea of running low power: 1 to 100 W output depending on frequency and other considerations. Despite ham radio's version of "old wives' tales" and other baloney, running low power—even with typical Stealth antennas—is *not* a big deal. You'll still make plenty of contacts, you'll still work DX and you'll still feel as though you're working from a real ham station—because you are.

Let me introduce you to the physics of signal strength. As you can see from the straight-line plot shown in **Figure 2.1**, a 1-W

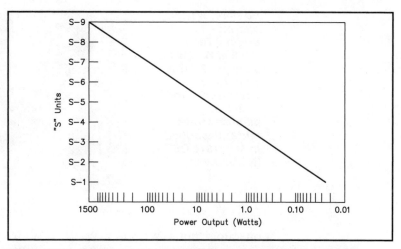

Figure 2.1—QRP is closer to QRO than you might think! This graph shows relative signal strengths for power outputs from 1500 W to 1 W. If your signal is 20 dB over S9 at 1500 W, it will be plenty strong at 5 W.

signal is almost an S-4, only a little more than 5 S-units weaker than a 1500-W signal. So, if your 1500-W signal is 20-dB over S9, your 1-W, 5-W or 50-W signal will be plenty strong.

Even if you didn't have to use Stealth Amateur Radio technology I'd still encourage you to run low power. Running low power, even at QRP levels, is a lot of fun. Just because you have a Stealth antenna, don't worry about low-power operating getting in the way of your ham radio enjoyment.

QRP

As long as we're talking about being resigned to low-power operating (by choice or circumstance), let's briefly look at QRP—the "official" pursuit of low-power ham radio. This is the kind of low-power operating that's done for the fun of it. By the way, the term *QRP* evolved from the CW procedural sign meaning, "I am reducing power," and *QRP?* "Shall I reduce power?"

So, how low is low? Well, a typical ham running 100 W output is putting out about 20 times more power than the 5-W CW output (10-W PEP output) that commonly defines "QRP power levels." But QRPers don't stop there. Some veteran low-power

Ed Hare, W1RFI, operates the original Tuna Tin 2 at W1AW in June 1999. (The Tuna Tin 2 was a two transistor — 350 mW — 40-meter transmitter described in May 1976 QST.) (N1RL photo)

operators run 1 W, 500 mW, 10 mW or even 1 mW of output power. "Microwatters," true enthusiasts (crazed individuals) who run less than 1 mW of output power, are a breed unto themselves! But make no mistake: They're out there!

Worldwide, QRPers number in the tens of thousands (or more), and you're more than welcome to join the ranks. Your comrades in spirit like nothing better than the challenge of working fellow hams while running just enough power to get through. Your 1-W signal will hardly dominate the band, but with the right conditions, you can easily work all 50 states and a *lot* of DX, even with Stealth antennas.

WORKING 'EM

Remember the friendly line in Figure 2.1? Your 1-W signal is only a little more than 3 S-units weaker than a 100-W signal. So, if your 100-W signal is S-9, your 1-W signal will be about S-6. That's plenty of signal! You'll listen more and call CQ less, perhaps, and persistence pays off, as does using the right approach. Beginning QRPers often call only the loudest stations. That's not necessary, although it's a good idea to have a good copy on the stations you do call.

Which bands to use? When the sunspot cycle is high (as it is at press time), 15, 10 and 6 meters (when it's open) are awesome, and stations with just about *any* kind of antennas can work the world. If you don't believe me, try it for yourself.

Twenty meters, of course, is the all-time bread-and-butter band—with lots of high-power competition. Forty and 30 meters are excellent bands for stateside QRPing. They can even deliver a fair amount of DX in evening and overnight hours, especially if you live near one coast or another. Thirty meters is favored by many QRP operators because it's quiet, uncrowded and "open for business" nearly 24 hours a day. Eighty meters is another good stateside QRP band; but it's not as popular as 40 meters because propagation is usually not as good (except for close-in contacts). Eighty also has DX potential, but competition is fierce and the physics of propagation are working against you. On MF—160 meters—QRP contacts are possible, especially when the band is quiet, but because the other HF bands offer much easier hunting, 160 can be a pretty lonely band for casual QRPers. **Table 2.1** lists the popular CW and SSB QRP calling frequencies on each band.

When it's time to get on the air, forget that you're running low power. After all, your signal is only a few S-units down from the big guns—but do let the other operators know that you're running low power. If you're tuning an uncrowded band, don't be afraid to call CQ. But do it like this:

CQ CQ CQ DE QRP NTØZ NTØZ

Table 2.1
QRP Calling Frequencies (MHz)

Band (m)	CW (MHz)	SSB (MHz)
160	1.810	1.910
80	3.560	3.985
	3.710 (Novice/Tech +)	
40	7.040	7.285
30	10.106	NA
20	14.060	14.285
17	18.096	18.130
15	21.060	21.385
	21.110 (Novice/Tech +)	
12	24.906	24.950
10	28.060	28.885
	28.110 (Novice/Tech +)	28.385 (Novice/Tech +)
6	50.080 - 50.100	50.125
2	144.060	144.285

When replying to a CQ, try:
W1XYZ DE QRP NT0Z NT0Z

If you get that "QRP" out there right away your response rate will soar.

COLLEGE QRP

I had my first taste of the truly amazing propagation that takes place at sunspot cycle maximums when I started college in 1980. I had access to the college club station, but because I lived off-campus and transportation (like 6-meter propagation) was sporadic, an apartment station was in order.

Not that I told my landlady, of course! One weekend when she was visiting her son in Minneapolis, I put up a "Stealth wire" which, by any standard, shouldn't have been expected to carve up the bands. The antenna was made from a 100-foot length of 28-gauge steel "picture hanging" wire. A 75-foot run spanned the house (between two modest trees), while the remaining 25 feet came straight down to a tent stake I'd pounded into the grass next to my window (it was a basement apartment). It was truly invisible unless I was standing next to the drop-down portion, which was in the backyard.

A fellow ham suggested the end insulators: 30-pound test monofilament fishing line tied to nearly clear plastic rings scavenged from a six-pack of diet soda. These worked so well that

The author used this Heathkit HW-8 QRP transceiver to work the world with a variety of Stealth antennas while in college. See the text for details. (author photo)

I've used them on many occasions since. More on that later.

This antenna essentially had no *real* RF ground. The "ground" consisted of a long aluminum knitting needle pushed into the ground next to the tent stake that held the vertical part of the pseudo inverted-L antenna. I fed the wire through a 15-foot length of RG-58 coax connected to a small MFJ antenna tuner.

I wasn't worried about RFI because my Heathkit HW-8 transceiver only put out a single watt of RF. I *was* worried, however, that I wouldn't make any contacts. What could one puny watt do while feeding such a pathetic antenna (it was all of 20 feet off the ground)? As it turns out—plenty.

In a few months of casual operating (ragchews and contests), I worked 40-some states and 30-some DXCC countries. Not bad for 1 W! Things really perked up when I upgraded to a Ten-Tec Argonaut transceiver, a classic 5-W, 80-10 meter QRP rig that works CW and SSB. (I wish I still had that little radio, and the Tempo One "QRO rig" I sold to pay for tuition one quarter!)

I vividly remember casually chatting with Scandinavian hams on 10-meter SSB. "Your Argonaut is doing fine business here, old man, you're 20 over S-9!" That Sunday morning, a whole string of European hams *called me*. Almost no one believed I was running low power!

When I moved to a newly constructed brick 16-plex apartment building (still on the basement floor), my knitting needle promptly went into the soil outside my ground-level bedroom window while the center conductor of the coax found itself "alligator clipped" to the bottom of a shiny new galvanized steel downspout. The building had four of them, all connected around the perimeter of the roof by a band of shiny steel. My "upside-down-four-square-with-single-wire-phasing" worked famously with either rig.

When I was forced to move into a fifth-floor dorm room, I dangled a 40-foot length of steel wire (weighted with a fishing sinker) out the window and connected the coax braid to the brick building's fire escape. Although DX stations were a bit more difficult to work with 1 W, even this compromise, upside-down vertical worked pretty well.

Had I known about counterpoises and RF grounding techniques at the time, these antennas would probably have worked even better. More on that later, too!

GEAR

Finding a rig for QRP work is pretty easy. There are many QRP-only rigs available, new and used. Look for Heathkit's long-popular HW series and Ten-Tec's Argonaut line-up. MFJ makes several single-band QRP CW transceivers, and if you're into kit building, check out Wilderness Radio's Sierra and the kits from S&S Engineering, Oak Hills Research and Ten-Tec. There are many others. Collectively, the ham magazines have published hundreds of home-brew "QRP stuff" in the past 10 years, so if you want to delve into "homemade radio," QRP is a good place to start.

If you don't want to invest in a dedicated QRP rig, it's relatively easy to reduce the power output of most modern solid-state rigs. The drive or power control can usually be used to trim the RF output to within acceptable QRP limits. Your rig's instruction manual will probably have more information.

QRP CLUBS AND AWARDS GALORE!

Many clubs exist to serve the interests of QRPers—and new ones seem to sprout weekly! One of the oldest and most prominent is the QRP Amateur Radio Club International (QRP ARCI). For information about QRP ARCI and a sample copy of its publication, *QRP Quarterly,* write to Mike Bryce, WB8VGE, PO Box 508, Massillon, OH 44648-0508. Other clubs include the Michigan QRP Club and the G-QRP Club, based in England.

Awards are very popular among QRP clubs and QRPers. QRP ARCI issues QRP versions of many popular awards (as does the ARRL) and several exclusive awards such as the 1000-mile-per-watt award. Contests are also popular among low-power enthusiasts. About a dozen QRP-only contests are held throughout the year, and many mainstream contests such as Sweepstakes, ARRL International DX, CQ Worldwide, and others have QRP entry classifications.

As for books on the subject, you need look no further than the ARRL Publications Catalog (or your favorite Amateur Radio products dealer)—it has several selections about QRP operating and QRP gear/construction. Check out *ARRL's Low Power Communication*, *QRP Power,* and *W1FB's QRP Notebook,* for starters. Several other publishers have also printed books about QRP. In addition you may want to watch the hamfest flea market

The author's 1986 QRP Amateur Radio Club International membership certificate. The man who signed it, former QRP ARCI General Secretary William Harding, K4AHK, worked DX and domestic stations by the thousands with low power and indoor antennas.

tables for copies of several ARRL publications that are no longer in print: *QRP Classics*, *Your QRP Operating Companion* and *Low Profile Amateur Radio*, to name a few.

HAVE A REAL CONVERSATION

Believe it or not, Stealth operation can actually boost your enjoyment of Amateur Radio. How? By forcing (helping?) you to get down to the nitty-gritty of one important aspect of ham radio—talking with other people (hams) about stuff that's mutually interesting.

With your Stealth station you'll make plenty of contacts, but you won't crack every pileup you jump into and you won't work every station you hear. Traditional pursuits such as DXing, chasing wallpaper and contesting are quite possible, of course, but between rapid-fire, robot-like QSOs (call sign, signal report, adios), why not have a few *real conversations*? So what if the other operator is in the next state instead of on some rare island?

There are millions of interesting individuals out there disguised as ham operators! Dig deeper—you won't be disappointed! By shifting your point of view you can find new enjoyment in Amateur Radio, regardless of band conditions.

TIPS FOR BETTER COMMUNICATING

Here are a few tips to "break the ice." Remember: Don't be shy! If necessary, just blurt something out. If your QSOs are stuck in a boring rut, dare to do something different!

- Get out a map or atlas. When you figure out where the other operator lives, check out his QTH on the map. That little blue squiggle might seem insignificant on your end, but your new friend might have been trout fishing there since he was a kid. By simply asking about the local geography, at least two things will happen: (1) you'll learn a lot more about that little blue squiggle (or whatever it is) and (2) you'll wake the ham on the other end up to the fact that a real conversation is about to take place. Both are big steps in the right direction.
- If you or your QSO partner live in a "famous place," feel free to get a little conversational mileage out of it. If you're chatting with someone in Winterset, Iowa, try out your best John Wayne accent. It couldn't hurt, could it?
- Asking people questions—on almost any topic—can often spice up an otherwise routine exchange. Be tactful, but ask away. Examples: "What do you do for a living?" "How about those Minnesota Vikings?" "Have you ever been to Japan?" You get the idea. To narrow down the range of possibilities, tailor your questions to what you already know about your QSO partner—or what you suspect.
- Within reason, feel free to let other hams know a little bit about what you're up to. Instead of keying the repeater with "This is W9XYZ, listening," try "This is W9XYZ, trying out my new totally invisible Stealth Station, listening." Which do you think would garner more responses on a typical sleepy repeater? Maybe the old-timer's CQ—"This is Bill, W9XYZ, calling CQ from the Louisiana bayou town of Swampy Creek"—heard regularly in decades past, has some merit. Don't use it while checking into an emergency net, and don't use it all the time, but you might give it a try on an uncrowded band just to see what happens.
- Be careful when discussing potentially controversial subjects such

as politics, religion, light beer, left-handed pitchers, etc. I'm not trying to step on your First Amendment rights. I'm merely suggesting that you be respectful and use common courtesy when bringing up certain topics. Amateur Radio *is* diverse, but it's also tolerant and accepting, and the best ham radio discussions build on a common ground of shared interests.

Regardless of which techniques you use (there are many more than those listed here), taking steps to make ham radio friends through better conversation will only increase your enjoyment of Stealth Amateur Radio. You never know when you'll make a lifelong friend you would have otherwise overlooked because of a "cut and dried" QSO!

STEALTH OPERATIONS

At this point I hope you're convinced that Stealth operators can do just about anything that can be done by most "average" radio amateurs. Sure, Stealthy moonbounce is unlikely, but it's just as unlikely for average hams. Working 160 meters with an attic antenna isn't likely to produce a contest superstation, but most hams who aren't restricted to Stealth techniques don't do all that well on 160 meters, either.

To maximize your Stealth experience, figure out the Amateur Radio activities you *can* do and focus on enjoying them. You can follow other paths later as your skills and situations improve. Be creative and keep exploring, but stay calm and move forward with reasonable expectations.

Working other stations from your Stealth shack can be a lot like making your first QSO. It's exciting! At first, when stations reply to your calls, you'll be amazed! As you gain experience you'll realize that your Stealth setup isn't so restrictive after all.

Before we move on, let's take a look at bands, modes and a few operating hints.

BAND BY BAND

As we take this Stealth tour of the bands, I'm assuming that you're using a rig that puts out 1 to 100 W and that you have some kind of Stealth antenna, indoors or out, and that the antenna is a compromise of one sort or another. The described activities and frequencies are far from comprehensive, so if you're not familiar with the ins and outs, get a copy of *The ARRL Operating Manual*, which goes into great detail.

Chuck Joseph, N5JED, now of Great Falls, Montana, worked WAC, WAS and DXCC while running 100 W to an MFJ Model 1621 portable antenna. The antenna, a 54-inch whip mounted atop a small tuning box, was mounted indoors on the second floor of a wood frame house in Mount Clemens, Michigan. If you ever get discouraged, look at this photo, imagine Chuck's tiny, inefficient antenna, and press on! (photo courtesy of N5JED)

Also, although I mention VHF/UHF operations in this book, I'm mostly concerned with 160 through 6 meters. The ham bands at 2 meters and up are inherently Stealthy, and if you learn how to be Stealthy on the lower bands, keeping a low profile on VHF and higher frequencies will be trivial!

160 AND 80 METERS

The low bands really reward Stealth operators who can erect outdoor antennas—the bigger the better! For example, working 80 meters with a typical *big* mobile antenna nets you an efficiency of between 2% (a poor antenna) and 8% (an excellent antenna). That's right, if you feed 100 W to an antenna that's 5% efficient, it radiates 5 W and wastes the rest on element heating and other useless (to us) phenomena. Depending on its design and overall length, your compromise Stealth antenna may be 50% efficient—or 1%. If your antenna isn't *resonant* at the operating frequency, it may present matching difficulties that may introduce additional losses. Keep that in mind!

Daytime QSOs are mostly local, but at night, contacts out to 1500 miles are common, even with Stealth antennas. Electrical noise—atmospheric and man-made—can be a real problem. QRPers hang out near 3560 kHz. Low-power operators are frequently better listeners, so don't be shy about calling them. On 75 meter SSB, "the ragchewer's subband," look for contacts from 3850 to 4000 kHz. Late night/early morning is my favorite time to work this part of the spectrum. On quiet (low noise) nights I often chat with the "gang" on 75 meters while running 10 W to a wire antenna.

40 METERS

Forty meters is a lovely all-around band for Stealth operators. During the daytime, expect solid QSOs out to several hundred miles. As late afternoon approaches, shortwave broadcasters dominate the band above 7100 kHz, making Stealth QSOs there rather difficult. The low end of the band from 7000 to 7100 kHz is loaded with stations from afternoon until the wee hours. Work the lower 48 states or add a DX QSO to your logbook. QRPers sling dits on 7030 and 7040 kHz. When conditions are favorable, 40 meters is a good DX band, especially if you live near the coast.

30 METERS

This band, arguably the best band for Stealth operators, is a lot like 40 meters—only better! Because power output for US stations on this band is limited to 200 W, interference from big signals is almost nonexistent. Thirty meters is open to somewhere all of the time, and the skip distances tend to be greater than those on 40 meters, making this a great Stealth DX band (CW and digital modes only). Thirty meters is a friendly, hassle-free band.

20 METERS

The first of the serious DX bands, 20 meters is crowded with DX stations and crowded with stations running big amplifiers—it's just plain crowded! At cooperative points in the sunspot cycle, however, 20 meters is open day and night, and an antenna for this band is much smaller and easier to conceal than its low-band cousins. The QRP calling frequency is 14,060 kHz, right in the "calmest" part of the band.

I don't really care for 20 meters. For years it's been my "band of last resort." I always check in on 20 meters during contests, but I've found that it's often less stressful to make and maintain QSOs on most other bands. Your mileage may vary, however!

17, 15 AND 12 METERS

Mostly used for daytime QSOs, these bands sport excellent Stealth potential. As with 20 meters, full-size antennas are physically small. At these frequencies, indoor and mobile whip antennas are reasonably efficient. When the band is open, worldwide QSOs are the norm. Stealth antenna? What Stealth antenna? SSB operation is a bit easier on 17 and 12 meters because these bands are relatively uncrowded.

10 METERS

When 10 meters opens during periods of high solar activity, it's the Stealth band of choice. If you're running 25 W, you'll have a big signal. Even during solar doldrums, 10 meters opens regularly via sporadic E and several other more exotic prop-agation modes. Look for SSB QSOs between 28.3 and 28.5 MHz. The QRP calling frequency is 28.060 MHz. FM is popular above 29 MHz.

6 METERS

Affectionately known as the Magic Band by enthusiasts, 6 meters is a "transition band" that bridges HF and VHF. Operating here can take a fair measure of patience, but during band openings, low power stations get out famously. Even when the band is "dead," local contacts out to 100 miles or so are common. At solar cycle peaks, 6 meters supports DX QSOs, and at other times, even during the doldrums, 6 meters regularly opens via sporadic E (especially in the summer months). Antennas, even multi-element directional antennas are physically small. In fact, a 6-meter beam antenna fits in many attics (or doubles as a disguised TV antenna, which may be allowed at your QTH).

2 METERS

What can I say about 2 meters? It's the most populated ham band in the US, and because the landscape is dotted with thousands of repeaters, you can work stations on FM almost around the clock, fixed or mobile. Two meters also sports packet radio activity, which can put you in touch with other hams in faraway places. SSB, CW and even satellite operations take place on 2 meters, and if you think antennas for 6 meters are small, 2-meter antennas will seem truly miniature. Directional antennas will easily fit in the attic or on the balcony. Store them in the closet when they're not in use.

1.25 METERS AND 70 CENTIMETERS

Much like 2 meters, 222 MHz is highlighted by the fact that it's one band that every US ham can use, regardless of license class. FM (voice, data, repeater and mobile) activities are predominant, but weak-signal operations (SSB, CW, moonbounce) can be found here.

In some parts of the country, 430 MHz is practically known as the Amateur Television Band. If you live in a metropolitan area, chances are good that a group of local hams is beaming fast-scan ATV signals (full-motion video and audio) back and forth, directly or via special repeaters. As you can imagine, antennas are quite small and easy to conceal.

If you have a "cable-ready" TV set, try connecting an antenna to the set and tuning Cable channels 57 through 61. If there is local ATV activity you may well be able to watch the signals!

HF MODES—WHICH IS BEST?

Speaking plainly, Morse code is the most effective mode for Stealth operators, bar none. I'm not being a CW snob, I'm just telling it like it is. I like voice modes as much as the next guy, but the data is irrefutable: CW gets through when voice modes fail.

Why? Well, for starters, CW signals require less bandwidth. Signals with economical bandwidths allow the use of similarly narrow IF filters. Narrow filters afford greater signal-to-noise ratios and provide excellent rejection of unwanted signals and noise. This all adds up to a tremendous benefit for operators running low power to compromise antennas (us!).

My recent experience on 160 meters is a perfect illustration. In about eight hours of casual contest operating I made 120 QSOs and worked 36 US states and four DXCC countries, including Alaska and Hawaii. I was running 100 W to a 40-meter loop—not exactly a killer low-band antenna. Granted, I had to use my rig's narrow CW filters *and* the CW filters in an external Timewave DSP box, but the QSOs went into the log. Working DX on 160 was a rush. I hadn't done that in 23 years as a ham. On SSB I made eight contacts, and most required seemingly endless call sign repeats. Many QSOs couldn't be completed or the other stations simply couldn't hear me at all. On CW, this happened only once or twice. The difference between the two modes isn't always this extreme, but then again …

You can still make Stealth SSB QSOs, of course, especially when the higher bands are open. I recommend using a good microphone and a correctly adjusted speech processor, if possible. Digital modes such as AMTOR (which uses error-correcting back-and-forth transmissions to compensate for low power and marginal copy) and PSK-31 (which uses *very* narrow filters to create a high signal-to-noise ratio) can boost Stealth success rates.

RESOURCES

For more information on the Amateur Radio Emergency Service (ARES) and the Radio Amateur Civil Emergency Service (RACES), contact your ARRL Section Emergency Coordinator (found in the Section News portion of *QST*). If you have Internet access, point your browser to **http://www.arrl.org/field/**.

For more information on SKYWARN, contact ARRL HQ or point your web browser to **http://www.skywarn.org**.

For information on Kachina computer-controlled (remote-controlled) transceivers, contact the manufacturer at PO Box 1949, Cottonwood, AZ 86326, USA; **http://www.kachina-az.com/**.

QRP CLUBS

- QRP Amateur Radio Club International (QRP ARCI), 848 Valbrook Ct, Lilburn, GA 30047-4280; **http://www.gqrp.com**
- G-QRP Club, US membership contact: Bill Kelsey, N8ET, 3521 Spring Lake Dr, Findlay, OH 45840; **http://www.btinternet.com/~g4wif/gqrp.htm**.
- Michigan QRP Club, 654 Georgia, Marysville, MI 48040; **http://home.online.no/~ analme/htmlrules/miqrp.html**

Chapter 3

Home is Where You Find Your Rig

For most hams, operating from a house, apartment, condo or dorm room is a Stealth Radio mainstay. Operating mobile or while vacationing are fine pursuits in their own right, but getting on the air from home—wherever or whatever that is—is what most operators crave.

Setting up an effective, Stealthy home station often requires some ingenuity. Every operating situation is unique in one way or another. There are a few guiding principles, however, and it's important to remember that there's always a way to accomplish your goals.

If you are one of the lucky few, you can operate from a spacious, well-appointed shack and use Stealth Radio techniques only for your antenna farm! If you're not, well, let's get started...

SITE SURVEY

Before you start drilling, sawing, assembling or stringing coax, examine your options and consider the positive and negative aspects of your various potential installations. Look at possible antenna locations, consider several shack sites, get a feel for how you'll accommodate the needs of nearby people — family members, roommates, neighbors and so on. Here are a few considerations.

SIZE AND LOCATION

Just where *can* you set up your shack, anyway? In the guest bedroom? In the den you paneled a few years ago—before the kids turned it into a playroom? How about the back porch? If you get rid of that old chest freezer you'll have plenty of room. What's going on in the basement? Basement shacks are a ham tradition, after all! Perhaps the closet under the stairs is big enough to hide a snug setup?

With a little creativity, you probably have more options than you might realize at first. Beyond the physical considerations, the needs of other people who share your living space must also be considered. Most people—nonhams and hams alike—don't appreciate a jumbled pile of gear hogging a corner of a "common area" of the house. Add a maze of tangled extension cords and an unsightly hole in the wall or ceiling and you have a recipe for disaster. So be discreet and be reasonable. What you can and can't get away with may ultimately depend on your negotiating skills!

BANDS AND MODES

To a certain extent, where you put your shack may depend on the bands and modes that interest you. We've already (hopefully) relegated the linear amplifier to the garage, but if you need to work the digital modes, for example, you'll need extra gear in the form of a computer or RTTY terminal. Although you might easily hide a laptop PC and a modern digital controller, concealing a desktop PC and a big ol' HAL RTTY terminal is something else entirely!

RADIOS AND ACCESSORIES

Speaking of RTTY terminals and computers, be sure to consider the space and storage needs of all necessary gear, including books, reference material, etc. Older rigs, although inexpensive, require a lot more desktop space. The same goes for antenna tuners, SWR/power meters, power supplies, clocks, lamps, keyers, and mics—the whole works.

Thanks to technological changes, we can fit an entire station, with every conceivable accessory, into the space formerly occupied by an older transceiver. Modern mobile rigs cover all modes from 160 meters to VHF and above. Many have built-in power meters, keyers, speech processors, narrow IF filters and digital signal processors. If you're cramped for space, or if you want a single rig for use at home, in the car or while on vacation, consider one of these mini marvels.

SECRET HIDING PLACES

If your station must truly be invisible, take a hint from the growing number of home office workers who manage to fit their entire offices—files, PCs, fax machines, etc—into an attractive cabinet. While closed, the setup looks like an armoire or a stand-alone closet. When you open it you have a compact, comfortable, fully equipped office. Although you probably don't need something that elaborate for your ham station, you can easily hide your gear in a roll-top desk, a cabinet, a bookcase that has a flip-up door, and so on. Use your imagination!

Figure 3.1— Although the author's present station doesn't have to be Stealthy, it's compact enough to qualify! (author photo)

Figure 3.2—Jim Kearman, KR1S, had to keep a compact, tidy—and hidden—HF station on more than a few occasions. In one apartment, Jim secreted his station in this armoire. He used the drawers to store headphones, logbooks and other goodies. (KR1S photo)

POWER TO THE PEOPLE

When scouting shack locations, be sure to consider available power sources. You probably won't need 240-V outlets (especially if the linear amplifier is still being stored in the garage), but a handy source of 120 V ac is probably necessary. Make sure the power requirements of your gear—add it *all* up—won't overload the circuit. Try to find a little-used circuit just to be on the safe side. If that's not possible, make sure your shack isn't sharing an ac circuit with high-power appliances such as heaters, air conditioners or microwave ovens. If you're interested in alternative power (especially good for QRP stations), you might be able to mount a solar panel on your windowsill and charge a sealed gel-cell battery to power your gear. Again, be creative, but don't string extension cords hither and thither.

FEED LINES AND ANTENNAS

If you have a choice, it's probably a good idea to set up your shack in a location that provides easy access to your antennas and/or feed lines. Coaxial cable is usually easier to string than twin lead or ladder line, but be sure to keep runs of antenna cable separate from ac power wiring—even coax.

Antenna wires or feed lines that run next to ac power wiring could result in electrical noise being picked up on the feed line. It could also result in the power wiring picking up RF and resulting in RFI to other devices. Finally, there is the danger that exposed wiring could come in contact with the antenna wires, resulting in a dangerous situation in which your rig becomes "hot" with ac, just waiting to shock you.

SOUND ADVICE

Even if our radios' dits, dahs, bleeps and squawks sound like music to *us*, others may find them truly annoying. Radio sounds might bother family members (usually solvable), but they might also "give your position away" to neighbors, fellow apartment dwellers or hotel guests (potentially disastrous). Remember the audio portion of your station operations during your shack-site selection process.

The sound factor is potentially troublesome for voice mode operators. For CW or digital operation you can put on a pair of headphones. The slight clicking of your keyer or the keys on your PC keyboard will go unnoticed. Your voice, however, even if you're whispering, can seem "megaphone powered"—especially if you forget yourself and are trying to "quietly shout" through the din of a pileup.

SAFETY FIRST!

As with any radio installation—Stealthy or otherwise—you must always consider safety issues. Amateur Radio isn't an inherently dangerous activity, but as hams we encounter electricity, batteries (sulfuric acid and hydrogen gas), physical danger (climbing towers, etc), RF radiation and more.

Death or injury doesn't result *only* from high voltages. Lower voltages can be just as deadly, if the current is large enough. (As little as 100 to 300 mA can disrupt your heart rhythm and cause death.) Working on rooftops and towers also calls for caution and common sense. By learning—and practicing—the right safety habits at the start of your ham career you'll hopefully avoid having to learn them the hard way. Like I did...

When I was a brand-new 13-year-old ham, my only rig was a TCS-6 AM/CW transmitter/receiver, a WWII-era boat anchor that I used on 80-meter CW. Although it worked fine, that grand old radio almost ended my then-short-lived ham radio career. It was my fault, actually, even though I thought I was being quite careful at the time.

The transmitter needed an adjustment, and I had it opened up—with the power on. I had placed the radio on a plywood workbench in the basement, near my operating position. As an added safety precaution I had placed a thick rubber insulating mat on the concrete floor (to stand on).

During the adjustment, as careful as I was, my foot edged off the mat and onto the bare floor. Somehow, because the screwdriver I was holding contacted a high-voltage source, or because of a grounding fault, a tremendous jolt of electricity slammed through my body.

A loud electric snap punctuated the fact that I had been thrown across the room! I hit the wall and crashed to the ground. The air had been expelled from my lungs and my heartbeat was faltering and irregular. Braaaap, braaap, it fluttered, bouncing around inside my chest. The room was spinning, and I thought I would soon be dead.

After an endless dozen seconds or so, my heartbeat returned to normal and my head started to clear. That incident—which forged a heightened respect for my own mortality—was a lesson I never forgot. Later, in college, I was excruciatingly careful as I built linear amplifiers and tube-type transmitters. My caution has paid off this far, and I have had no further accidents.

Other hams haven't been as lucky...

In the mid-1980s, an experienced North Dakota ham was killed when a vertical antenna he was installing accidentally touched an overhead power line. In the late 1980s a ham from Texas—with thousands of hours behind the key and test bench—was fatally shocked when he touched a high-voltage line inside his linear amplifier. Had he lived he would have had to adjust to the fact that the powerful jolt had *charred his hands completely off his body*. While operating Field Day-type stations, hams have electrocuted themselves by running power cords (plugged into gas-operated generators) through standing water. More than a few hams have also been killed by lightning strikes.

I'm not trying to be morbid—I just want you to keep these tips in mind while building, repairing, installing, adjusting and operating your Amateur Radio equipment.

CLIMBING SAFETY

- Never climb alone—and that includes towers or ladders. Always use a helper/spotter.
- When working on a tower, always wear and use an approved, secure safety belt.
- Plan your work before you start. Have the proper tools and materials on hand.
- Take a break every now and then.
- If you're uncomfortable working at heights, stay on the ground and get help from an experienced climber.
- Stay away from—and be alert for—power lines or other overhead wires.
- Don't climb when you are tired or distracted.

ELECTRICAL SAFETY

- Make sure the ac outlets in your shack are correctly wired and properly grounded.
- *Personally* disconnect equipment from power sources before connecting or adjusting them.
- Drain (ground) electrolytic capacitors before touching them.
- Don't work alone.
- Use tools with insulated handles.
- Install a master "power cut-off switch" in your shack or near your test bench and make sure everyone in your household knows how to use it.
- Work in illuminated areas.
- If you must service equipment while the power is on, follow the electrician's example: Keep one hand in your pocket while you work—that way, electrical energy won't have an easy path across your chest (and through your heart!) should your working hand contact a live source.

RF EXPOSURE

The **RF Exposure** sidebar in Chapter 1 provides some basic

RF AWARENESS GUIDELINES

These guidelines were developed by the ARRL RF Safety Committee, based on FCC/EPA measurements and other data.

- Although antennas on towers (well away from people) pose no exposure problem, make certain that the RF radiation is confined to the antenna radiating elements themselves. Provide a single, good station ground (earth), and eliminate radiation from transmission lines. Use good coaxial cable or other feed line properly. Avoid serious imbalance in your antenna system and feed line. For high-powered installations, avoid end fed antennas that come directly into the transmitter area near the operator.
- No person should ever be near any transmitting antenna while it is in use. This is especially true for mobile or ground-mounted vertical antennas. Avoid transmitting with more than 25 W in a VHF mobile installation unless it is possible to first measure the RF fields inside the vehicle. At the 1-kW level, both HF and VHF directional antennas should be at least 35 ft above inhabited areas. Avoid using indoor and attic-mounted antennas if at all possible. If open-wire feeders are used, ensure that it is not possible for people (or animals) to come into accidental contact with the feed line.
- Don't operate high-power amplifiers with the covers removed, especially at VHF/UHF.
- In the UHF/SHF region, never look into the open end of an activated length of waveguide or microwave feed-horn antenna or point it toward anyone. (If you do, you may be exposing your eyes to more than the maximum permissible exposure level of RF radiation.) Never point a high-gain, narrow-beamwidth antenna (a paraboloid, for instance) toward people. Use caution in aiming an EME (moonbounce) array toward the horizon; EME arrays may deliver an effective radiated power of 250,000 W or more.
- With hand-held transceivers, keep the antenna away from your head and use the lowest power possible to maintain communications. Use a separate microphone and hold the rig as far away from you as possible. This will reduce your exposure to the RF energy.
- Don't work on antennas that have RF power applied.
- Don't stand or sit close to a power supply or linear amplifier when the ac power is turned on. Stay at least 24 inches away from power transformers, electrical fans and other sources of high-level 60-Hz magnetic fields.

information about the concerns of exposure to radio frequency fields, and the FCC rules for hams to follow when installing, testing and operating their stations. Complying with the regulations isn't difficult. In fact, the precautionary measures are mostly common sense.

The ARRL RF Safety Committee has written much about the issues and research surrounding RF exposure safety. Every active ham should be familiar with the ARRL publication, *RF Exposure and You,* as well as the material in the latest edition of *The ARRL Handbook.* For a brief summary of some important guidelines, see the **RF Awareness Guidelines** sidebar. You can also find the full text of the *Handbook* RF Safety material and additional resources at **http://www.arrl.org/news/rfsafety/**.

In addition, the FCC's Office of Engineering and Technology has written *OET Bulletin 65* (the FCC Bulletin detailing RF exposure rules for various radio services) and *Supplement B to OET Bulletin 65* (concerned with the Amateur Radio Service). These documents are both included in ARRL's *RF Exposure and You.* The FCC's OET web site, **www.fcc.gov/oet/info/documents/bulletins/#65** has *OET Bulletin 65* and *Supplement B to OET Bulletin 65* available for viewing and downloading. Copies can also be purchased from International Transcription Service, Inc, 1231 20th Street, NW, Washington, DC 20036; telephone 202-857-3800; fax: 202-857-3805.

Ham radio books—including this one—often seem "preachy" when it comes to staying safe. By now, I'm sure you know why! Common sense and clear thinking cover almost every situation. Have fun...and be safe!

Chapter 4

Antenna Systems and Components

This chapter, and the one that follows, make up an important part of this book. As we discussed earlier, antennas are the key to ham radio success—Stealthy and otherwise. We'll cover a lot of material, but when we're done, you'll know how to put up an effective antenna for just about any situation. I hope you'll also come away with a better understanding of some potentially confusing antenna system concepts.

Before we begin, let's define our starting point:
- To save enough pages to fill a companion volume, I'm assuming that you have a basic knowledge of how to build wire Amateur Radio antennas, how to attach center and end insulators, how to solder coax or open-wire line to an antenna's feed point, and so on. If you've never done this stuff before you can find excellent tutorials in *Your Ham Antenna Companion*, by Paul Danzer, N1II, and a recent edition of the *ARRL Antenna Book* (both are available from the ARRL). In addition to looking over the books, try to find a ham radio buddy to help you with an antenna or two. Once you have even a little experience, constructing the simple antennas we're working with here will be well within your comfort zone.
- I'm also assuming that you're familiar with a few basic ham

antennas—dipoles, loops, verticals and random-length wires. You don't have to know the physics behind how each antenna works, but being able to recognize the various types and know a little bit about them will be helpful. The antennas we'll work with here are simple, simple and simple—nothing *too* unconventional!

- Although I have built dozens of antennas over the years, I am not an engineer. I can't do much with a Smith Chart (other than use it as a coaster for a hot drink), and I can't explain the physics and math behind *why* an antenna works. I can, however, show you *how to build* a bunch of decent Stealth antennas. I've built them myself and used them on the air. I know that they work, and that's enough for me. If you want more information or scientific confirmation (and you just can't have *too much* info when it comes to antennas), check out the antenna books offered by the ARRL and the RSGB. They're all available from *ARRLWeb* (**www.arrl.org**) or see the listings in any issue of *QST*.

Before we get to the designs themselves, let's review the basics of antenna hardware, feed lines, grounding, counterpoises, tuners, baluns, insulators and support ropes/lines. These are important parts of any antenna system, but they become even more important when we're building Stealth antennas, which often exist in less-than-ideal situations.

PHILOSOPHY 101

Hams have been designing, building and arguing about antennas since the earliest days of radio. After zillions of antenna articles, hundreds of books and a near-infinite number of on-air discussions, antennas remain a mystery to many and a topic of hot debate among most amateurs. What works for one ham might fizzle for another. Antennas that meet every possible scientific test in the laboratory might—or might not—work in the field. Antennas that can't possibly work—like bumblebees that *can't* fly—sometimes surprise us. Every installation has unique qualities, and trial and error is sometimes the only way to come up with an effective antenna.

GOOCH'S PARADOX

If there's one encompassing statement about home-brewed antennas, it's summed up in Gooch's Paradox: "RF Gotta Go

Somewhere." This unquestionable truism was handed down to me by Dave Newkirk, W9VES, a former HQ staffer and all-around radio mentor. Gooch's Paradox is often invoked when explaining the performance—good or bad—of a particular antenna. Think about it. "RF Gotta Go Somewhere."

In fact, RF might:
- Flow into your antenna (and become useful, radiated RF energy).
- Heat up your coax (not as useful).
- Radiate from your station's alleged ground connection (probably not good).
- Interfere with your neighbor's TV (you decide).
- Be rectified by the rusty connections that sometimes form between rain gutters and downspouts (amazingly bad).
- Feed back along the outside shield of your antenna lead and burn your lip when you speak into your microphone (zap!).

RF *might* go just about anywhere, but it's gotta go somewhere. In this chapter we're concerned with getting RF to flow into your antenna—whatever your antenna may be—where it can be radiated into space. In Chapter 6 we'll deal with RF that goes elsewhere!

THE BIG THREE

Here are a few cornerstones of Stealth Antenna Design:
- *Higher is better*. In general, the higher an antenna is, the better it performs. This isn't *always* true, of course, but for our purposes it's pretty much a fact.
- *Bigger is better*. In most situations, the bigger an antenna is (in length and wire/element diameter), the better it performs. A 20-foot vertical whip works better than a 10-footer, which works better than a five-footer. If you could make a full-size dipole from solid copper wire the diameter of a sewer pipe, it would work better than a similarly sized dipole made from garden variety antenna wire.

Taken to extremes, however, this rule falls apart. A horizontal loop antenna made with 20 miles of wire probably won't work better than a loop cut for 80 meters. In fact, it might not do much of anything! For the most part, 20-mile-long Stealth antennas aren't common enough to worry about (Texas ranch hams excluded).

- *Outside is better than inside.* With a few exceptions, outdoor Stealth antennas generally outperform indoor antennas. Although a 10-meter dipole inside the penthouse suite of a skyscraper may work better than a similar outdoor dipole 10 feet off the ground, most Stealth operators will want to try to put up some kind of outdoor antenna if at all possible.

GROUNDS

The various kinds of grounds—and what exactly constitutes a good ground connection—are often confusing and mysterious. As hams we suspect that a good ground is necessary for lightning protection and for good antenna performance, but we might have only vague ideas about what these things really are and how we might achieve them.

After all, we read or hear about ac and dc grounds, ground planes, RF grounds, grounded wall outlets, grounded electrical plugs, ground-fault circuit interrupters, Earth grounds, Earth connections, grounded wrist straps (for working with sensitive semiconductors), grounded grid amplifiers, grounded this and grounded that. What is all this grounding stuff, anyway? As hams, we need to be primarily concerned with two types of grounds: electrical (safety) and RF.

If you are familiar with residential house wiring, or if you carefully examine the electrical box that contains your house's circuit breakers or fuses, you'll see that the ground terminal of each electrical outlet in your house is connected to a common wire that runs from the junction box to a copper rod that's been driven into the ground outside your house. This ground rod may also be electrically connected to buried water pipes or other underground metal. This is an electrical safety ground. It's designed to protect you (and your house) from electrical shock and related fire hazards. It's probably a poor RF ground. If you connect the RF portion—the antenna system—of your ham station to it, you may cause or aggravate potential interference and noise problems. In most cases, your station is already connected to this *safety ground* through a properly wired 120-V outlet.

An RF ground—and exactly what it does—is sometimes more difficult to understand. I often think of an RF ground as a giant sheet of aluminum foil that covers the surface of the earth (purists and technical types should play along with me here for a bit). If my radio station has a good connection to the foil, my antennas can bounce their transmitted signals off the foil, increasing their performance and reducing the potential for some kinds of interference.

The *radio-related* functions of an *RF ground* are distinct and separate from the *safety related* functions of an *electrical safety ground*.

To complicate this clear statement, it's important to know that although every station needs an electrical safety ground, some stations may work better and may cause less interference by eliminating conventional RF grounds!

BALANCED VERSUS UNBALANCED ANTENNAS

How can that be? In many ways, the difference between balanced and unbalanced antennas is responsible. Again, using our simplified aluminum foil RF ground, most unbalanced antennas (verticals and end-fed wires, for example) require a good connection to the foil (the RF ground) to function efficiently. Balanced antennas (dipoles and loops, for example), on the other hand, don't need an RF ground connection to function correctly. Essentially, because of their electrical characteristics, they have built-in RF grounds.

Explaining the precise theory is beyond the scope of this book. Try to think of it by looking at a 40-meter dipole. It's made from two wire legs, each 33 feet long, and each running roughly horizontal (parallel with the ground). A 40-meter vertical, on the other hand, consists of a single 33-foot vertical element mounted at ground level, but electrically insulated from the ground.

When we connect a coaxial feed line to the dipole, the center lead is connected to one wire leg while the coax shield is connected to the other. See **Figure 4.1A**. When we hook up the vertical, the center lead is connected to the vertical element, while the coax shield is connected to the RF ground, or the foil. See Figure 4.1B.

The foil—the RF ground—makes up one half of the vertical antenna! Without a good connection to this RF ground, the vertical antenna won't work very well. From Gooch's Paradox we know that RF Gotta Go Somewhere, and if you break a vertical antenna's RF ground connection—or if you disconnect the coax

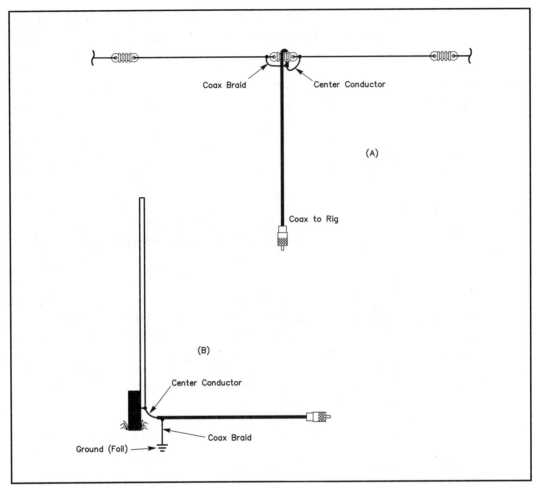

Figure 4.1—Coax connections for typical dipole and vertical antennas.

shield leg of a dipole antenna—RF goes *somewhere*, but most of it doesn't get radiated into space.

Ignoring that such an antenna would probably exhibit a very high SWR, and that your rig may not provide much power to such a load, the RF that did go somewhere would almost surely be minimal. The chassis of your rig is connected to a safety ground via the house wiring, however, so the RF energy might seek a connection to the foil through the safety ground, flooding your house with unwanted RF, potentially causing interference and other unwanted anomalies.

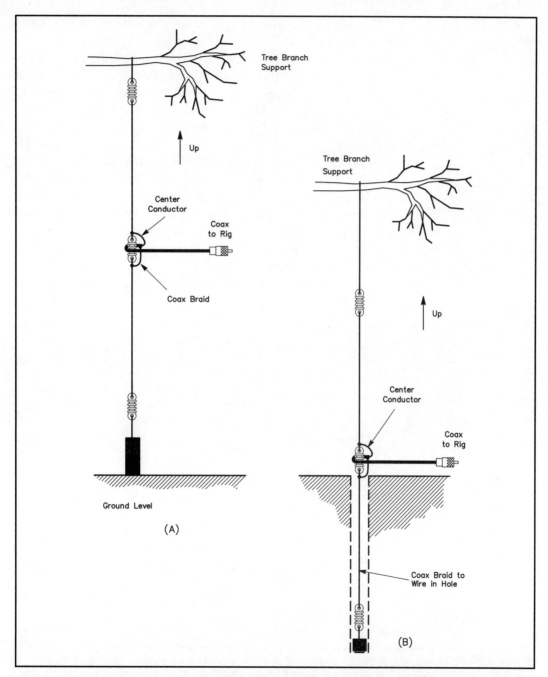

Figure 4.2—A vertical dipole (A) would look a lot like a ground-mounted vertical if we could drop the lower leg into a properly sized hole (B).

I hope this RF ground stuff is starting to make sense. Let's go one step further.

If we used a well-drilling rig to bore a 33-foot deep hole in the backyard, we could take our 40-meter dipole antenna, turn it so that the wire elements were straight up and down (a vertical dipole, as shown in **Figure 4.2A**), and lower the leg that's connected to the coax shield into the hole (Figure 4.2B). Hey, that looks a lot like a 40-meter vertical! So, the ground—the RF ground—really does make up half of a vertical antenna!

As it turns out, most ground—soil—doesn't make a very good RF ground. To help our 40-meter vertical antenna to radiate a better signal, we could take one or more 33-foot-long pieces of wire, connect them to the shield of our coaxial feed line and lay them on top of the soil, fanning them out from the base of the vertical element (**Figure 4.3**).

These, of course, are *ground radials*. Because soil is usually a poor RF ground, we can use wire (or, if we're feeling

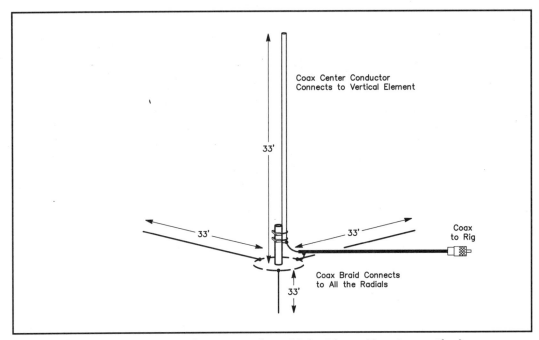

Figure 4.3—To improve the performance of our 33-foot-long 40-meter vertical, we can place three 33-foot-long radials on the ground (or slightly below the surface). The center conductor of the coaxial feed line is connected to the vertical element while the coax braid is connected to all three radials.

extravagant, an acre or two of copper foil) to make an RF ground. If we used a single wire to make a ground radial, our antenna would look like an L-shaped dipole with one vertical element and one horizontal element. In keeping with our simplified model, one leg is the antenna, while the other leg is the RF ground.

If we hoist the whole thing into the air, we have an elevated L-shaped dipole (**Figure 4.4**). Even if we move the ground leg

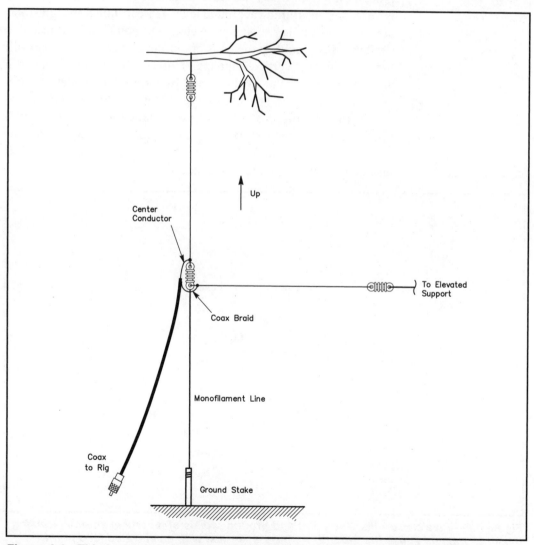

Figure 4.4—This L-shaped elevated vertical looks a lot like an L-shaped dipole.

Antenna Systems and Components   4-9

back into its traditional dipole position, one leg is still the antenna, while the other leg remains the RF ground. That's why I can get away with saying that dipoles (as well as loops and other balanced antennas) have built-in RF grounds. That is also why verticals (along with end-fed wires and other unbalanced antennas) need good RF ground connections.

By the way, if we lay out three ground radials to work as an RF ground for our 40-meter vertical antenna, there is no reason why we have to keep everything at ground level. If we installed an eight-foot-tall fence post to serve as an elevated feed point, and we installed similar posts to hold up the ends of our three radials, we would still have a 40-meter vertical antenna. This con-figuration is called a ground-plane antenna (widely used by CBers on 11 meters) or a vertical antenna with an elevated feed point. See **Figure 4.5**.

In fact, lifting the antenna and radials above the soil level often improves its performance! As it turns out, with ground-mounted verticals, some RF energy is wasted in heating the soil instead of being radiated into space.

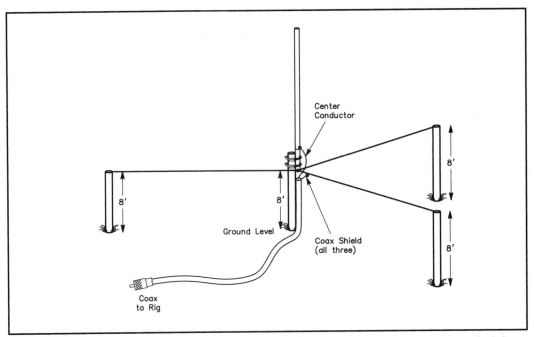

Figure 4.5—If we elevate the feed point and ground radials of a vertical antenna (raising them several feet above the ground) we have a vertical with an elevated feed point, commonly called a ground-plane antenna.

WHERE IS RF GROUND, ANYWAY?

Knowing that elevated verticals often work better than ground-mounted, traditional verticals raises an interesting question. Where is RF ground, anyway? Is it at the surface of the soil? Fourteen inches below the deepest grass roots? A mile down? Where is RF ground in relation to mountaintops or skyscrapers? If soil doesn't always make a good RF ground, what about solid rock? Or saltwater? Or the trunk lid of a car? The point here is that RF ground is where we make it!

Have you ever wondered why those lucky hams who can manage to operate from the beach of a South Pacific island always seem to have such strong signals from their beach- or surf-mounted verticals? Saltwater, unlike most soil, is a fabulous RF ground. So, at the ocean's edge, anyway, RF ground is wherever the saltwater is.

In our backyards—real ground as antenna modelers like to call it—RF ground may be near the surface, or it may be a mile underground. At the peak of a Himalayan mountain, RF ground may be five miles underfoot. In your apartment building, who knows where RF ground is? If we make our own RF ground, however, we always know where it is.

THE VERSATILE COUNTERPOISE

Let's say we're using an end-fed wire that runs from a second-story window to a tree in the backyard. Well, for RF ground purposes, that's just a vertical antenna turned on its side. An end-fed wire (a side-mounted vertical in disguise) needs an RF ground to make up the other half of the antenna system. We don't have access to the real RF ground, so we need a universal RF ground kit to complete our installation. That kit is called a counterpoise. You'll recognize it as a ground radial in disguise!

Let's say we've cut a 33-foot (quarter wavelength) end-fed wire to work on 40 meters. If you think it's logical to cut a 33-foot counterpoise to serve as an RF ground, you're absolutely right! Simply connect the backyard wire to the output of your transmitter and the counterpoise to the transmitter chassis and you're set!

Construct the counterpoise from *insulated* wire and run it around the baseboards of your room. Wrap a piece of electrical tape on the far end to insulate the exposed end—high RF voltages can develop there, and you don't want to shock anyone, including yourself. See **Figure 4.6**.

If you're using an antenna tuner to load the backyard wire on several bands, simply cut a quarter-wavelength counterpoise wire for each band, connect them together at the antenna tuner end and connect the mass of ground radials to the tuner's

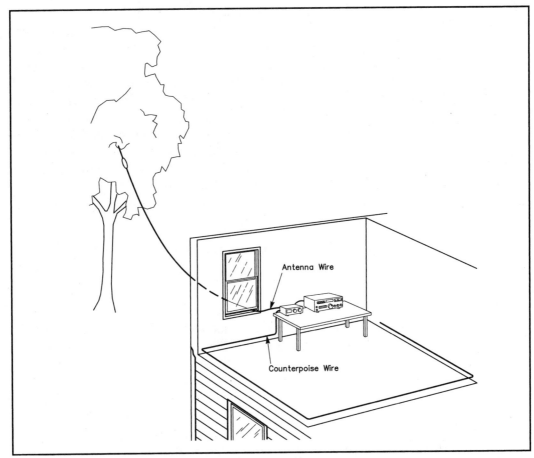

Figure 4.6—This 33-foot-long end-fed wire (a common Stealth antenna) is really a 40-meter vertical antenna turned on its side. The other half of the antenna, the RF ground, is provided by the 33-foot-long counterpoise. Connect one end of the counterpoise to the transmitter or antenna tuner chassis and string the remaining length of wire around the baseboards of the room. Both wires are a quarter-wavelength long at the chosen operating frequency, which makes them look (and work) like the two quarter-wavelength legs of a dipole!

grounding post. Some operators use a single counterpoise that's a quarter-wavelength long at the lowest band of operation. This works, too.

A counterpoise not only improves the performance of your antenna, it often keeps RF energy from flowing through circuits where it shouldn't, often eliminating pesky RF in the shack and RF interference problems.

ARTIFICIAL GROUNDS

Sometimes, no matter what you do, achieving a good-performing RF ground—especially in stations that are on the second floor or higher—seems impossible. Counterpoises, the Stealth operator's mainstay, don't resolve *every* RF grounding situation. They may make antennas load and radiate more efficiently, but feed line imbalance or other factors may cause RF feedback in the shack, RFI, and so on. One way to improve difficult installations is to use an "artificial ground"—an antenna tuner for your ground connection or counterpoise. Artificial grounds are really "counterpoise tuners."

MFJ's Model 931 Artificial Ground, shown in **Figure 4.7**, is a popular commercial "ground tuner." A similar unit is available from Ten-Tec. Both units use a simple L-network with a tapped coil and a variable capacitor. Connect the Artificial Ground between your rig's chassis and your counterpoise (or other ground) and tune the network for maximum current as shown on the built-in meter.

Some operators wouldn't be without their counterpoise tuners. I've used the MFJ unit to eliminate problems with RF in the shack on 15 meters. They can be lifesavers, but they're not a cure-all. We'll look at artificial grounds again in Chapter 6.

SOMETHING TO WORK AGAINST

Now that we know how to use a counterpoise, that a counterpoise is a portable RF ground, and that a counterpoise works as the other half of an unbalanced antenna, you might be wondering whether other "things" can be used as RF grounds.

If you've ever worked a mobile station you already know the answer. Because most mobile operators don't trail quarter-wavelength pieces of wire behind them as they travel down the

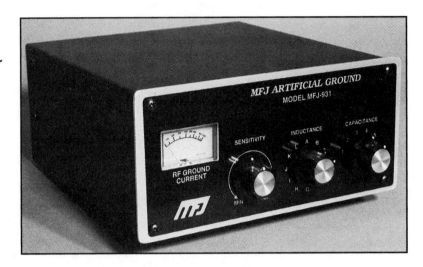

Figure 4.7—MFJ's Model 931 Artificial Ground is really a counter-poise tuner that can often help Stealth operators achieve more-effective RF grounds. (author photo)

highway, they must use something else as an RF ground. That something is the metal body of the car. This is called "working against" the metal car body. A 40-meter vertical works against its radial system, an end-fed wire works against its counterpoise and a mobile whip works against the metal body of a car.

Some people have good luck working wires or whips against balcony railings or metal fire escapes. Some people even stick mag-mount 2-meter whips on top of refrigerators—or cookie sheets or metal bedpans! *Sometimes*, working a whip or a wire against a large metal object makes for a great Stealth antenna.

FEED LINES, TUNERS AND BALUNS

Since WWII, hams have been seduced by the convenience of 50-Ω coaxial cable. It's easy to use, affordable, waterproof, buriable, shielded (at least in theory!), flexible and available in a wide variety of sizes and power-handling capacities. Most hams think it's probably the best feed line available and, for some situations, it's just fine. For others, however, it's *horrible*, and for some, it's downright *disastrous*! The problem usually shows up when we try to feed a single antenna on multiple bands.

The traditional multiband dipole is fed with a random length of 50-Ω coax that's tweaked into submission by an antenna tuner. Many "experts" tell you to put up as much wire as possible and let the tuner worry about matching the load on various bands. Even on bands where the antenna's SWR is quite high, and a lot of energy is reflected back and forth between the tuner and the antenna, *some* RF energy will be radiated.

We know from Gooch's Paradox that RF gotta go somewhere, but it doesn't have to go *anywhere* in an elegant and useful fashion! In the high SWR conditions often found in typical multiband, tuner-fed dipoles, Gooch's Paradox might as well read, RF gotta heat the feed line!

Wait a minute! We've always been told that antenna tuners enable us to use non-resonant antennas on several bands without a hitch. What gives? Actually, although antenna tuners can match a wide variety of antenna impedances to the 50-Ω output of your transmitter—so your rig will effectively put out its rated power—if the SWR on the coax between the antenna and the tuner is very high, little power will be radiated by the antenna even if everything seems matched.

For example, a 66-foot non-resonant dipole fed with 50 feet of high-quality, low-loss coaxial cable will tune up on all bands, 40 through 10 meters. Tuning on some bands will be touchy, but you can work stations, DX included. How much power is being wasted because of high SWR, though?

The manufacturer's data sheet says our coaxial cable has 1.5 dB of loss per 100 feet at 100 MHz (loss increases with cable length and frequency). We're using only 50 feet with an upper frequency limit of 30 MHz, so our losses due to SWR mismatches should be minimal, right?

Wrong. Those loss figures are for matched, resonant antennas. With high SWR values, a lot of power (sometimes *most* of your power) can be lost between your antenna and tuner—even with a low SWR between your rig and tuner. As we'll see, losses increase in proportion to SWR, too. A 3-dB loss represents a 50% reduction in transmitted signal strength!

On 40 meters, our 66-foot dipole is a great match, and the antenna system wastes only about 0.2 dB. Not bad! On 15 meters, an odd harmonic of 40 meters, the match is also pretty good, sporting an acceptable 0.8 dB loss. What happens if we try to load this antenna on the lower-frequency bands, though? On 80 meters, feed line losses approach 14 dB, and on 160 meters, losses total a staggering 27 dB![1] If we start with a typical 100-W output, we'll radiate about 3 W on

Table 4-1

Loss Figures for Common 50-Ohm Coaxial Cables

Cable	Loss in Decibels (per 100 feet at 100 MHz)
RG-174/U	8.4
RG-58	4.2
RG-58/U	3.8 - 4.5
RG-8/X	3.4 - 3.7
RG-59	2.9
RG-8A/U	2.8
RG-59/U	2.5 - 4.0
RG-6	2.3
RG-213/U	2.1
RG-6/U	2.0 - 2.1
RG-11	1.4
RG-8/U	1.3 - 1.9
RG-11/U	1.2 - 2.0
9913	1.3
LMR-400	1.2
LMR-600	0.9
LMR-900	0.6

80 meters and less than a half a watt on 160! No wonder your mileage may vary!

Typical line losses for various types of coaxial cables used in Amateur Radio applications are shown in **Table 4-1**.

SOLUTION 1

One way to reduce the feed line losses experienced while using multiband or non-resonant antennas is to ditch our traditional coaxial feed line and replace it with 450-Ω ladder line (**Figure 4.8**), or open-wire line—which is even more traditional!

As shown in **Figure 4.9**, 450-Ω ladder line—parallel conductors separated by a plastic, ladder-like insulating material—replaces the coax we previously used to feed our dipole. (One common variation of this type of 450-Ω line is often called "window line" because it consists of two conductors separated by about a 1-inch wide plastic ribbon, with square "windows" cut out of the line every inch or so.) Ladder line, also known as 450-Ω balanced line, was the norm in the days before coaxial cable (an unbalanced line). It may not be as convenient as coaxial cable, but when used with an antenna tuner designed to handle ladder line (most can do at least an okay job), feed line losses for our 66-foot dipole stay blissfully below 0.3 dB on all bands, 40 through 10 meters! On 80 and 160 meters—big trouble spots when fed with coax—losses total 1.4 and 8.6 dB, respectively. That's a *tremendous* improvement!

If ladder line were a magic cure-all, of course, we'd never use coax. As it turns out, open-wire line has a few quirks of its own:
• When attaching balanced feeders to houses, structures and towers, be sure to keep the ladder line several inches away from metal (or metal-containing) objects. Running the ladder line near large metal objects can unbalance the balanced feeder, which can lead to reduced performance, RFI and RF in the shack. Stand-off

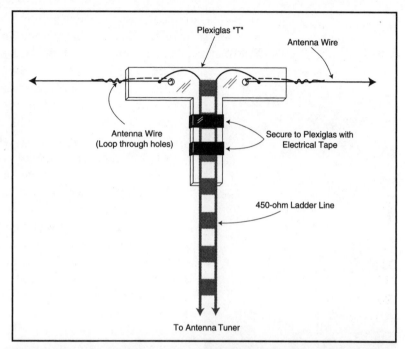

Figure 4.8—This multiband dipole is fed with 450-Ω ladder line and requires a Plexiglas center insulator (or something similar) that's shaped to reinforce the line to keep it from flexing and fatiguing.

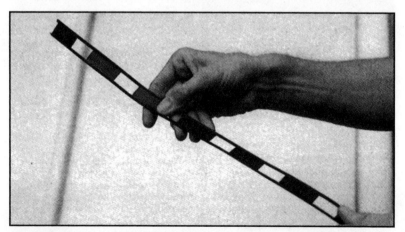

Figure 4.9—The most common type of 450-Ω ladder line uses plastic insulating material to maintain a constant separation between the two conductors. The gaps in the insulation give the line its ladder-like appearance. This type of commercially constructed line is sometimes called "window line."

Antenna Systems and Components   4-17

insulators commonly used when installing TV antennas that are fed with 300-Ω twin-lead are available at RadioShack stores nationwide. You can also make your own. I often use thin sticks of scrap Plexiglas or scraps of heavy plastic food-storage containers bolted onto small L brackets (**Figure 4.10**).

- Be sure your antenna tuner has a sufficient voltage rating. Tuning antennas with a high feed line SWR can create *very* high RF voltages inside your tuner. Resulting arcs and sparks can damage expensive equipment (especially on bands with the highest SWR). Low-power operators have the advantage here.
- If arcing occurs, reduce your transmitter power output or get a tuner with beefy components. Using a 1-kW tuner (with bal-anced outputs) with your 100-W transceiver *isn't* excessive.
- Water, ice and snow can affect (unbalance) ladder line. Keep things clear for best results.
- If left flapping in the breeze, the soldered connection between your ladder line feeders and your dipole wires will probably fatigue and break rather quickly. Be sure to reinforce the junction with electrical tape as shown in Figure 4.8.
- Ladder line can be hard to find in some locales. If your local ham store doesn't stock it, check the ham magazines for wire and cable suppliers. Some operators—especially QRPers—sometimes use 300-Ω TV twin-lead instead of 450-Ω line. It's a true balanced line, but the reduced spacing between the wires and less-stringent insulation doesn't *always* produce acceptable results. If you're running low power, try it and see for yourself. I've successfully fed several QRP antennas with TV twin-lead.

If you're suffering from antenna restrictions of any type, a balanced feed line can provide an excellent compromise between convenience and cost. Simply install the longest center-fed dipole (or loop) that's practical (make each side the same length) and feed it with enough ladder line to comfortably reach your station. Don't worry about feed line length. Some hams

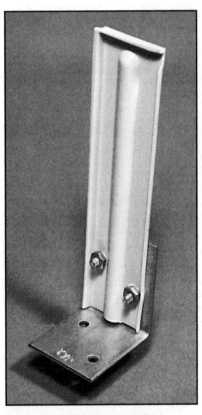

Figure 4.10—Stand-off insulators are used to keep runs of ladder line away from metal, rooftops and walls.
Commercially available stand-off insulators used for 300-Ω TV twinlead are available at any RadioShack store. Shown here is my homemade version made from heavy plastic food-storage container scraps and small "L" brackets. To make things easy I simply tape the ladder line to the end of the plastic. It's not elegant, but it's inexpensive and effective! (author photo)

use 500-foot runs of 450-Ω line and laugh at the losses (which, when installed and matched correctly, are practically microscopic). With a decent tuner (the beefier the better), you'll put out a good signal on a variety of bands.

SOLUTION 2

Sometimes, open-wire line isn't an option. Because balanced lines don't usually like to be fastened to metal buildings, trees, icy roofs, conduits, the ground, and so on (they like to hang in the air!), coaxial feeds are sometimes necessary. Don't we have to worry about the SWR, though? In typical antenna tuner installations, the cable that runs from your rig to the antenna tuner sees a low SWR. Most of the power transferred through it is passed on. The cable that runs between the tuner and the antenna is the part that has the potential SWR losses—even if the tuner can match the antenna. In high SWR conditions, much of the power transferred through that leg is wasted. (See **Figure 4.11A**.)

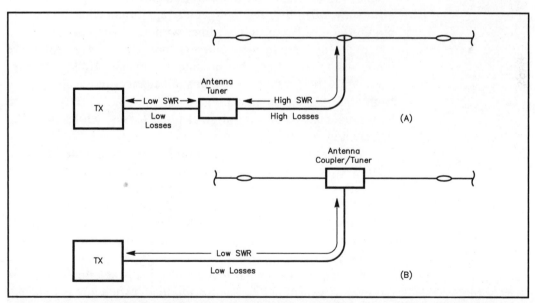

Figure 4.11—When feeding a typical multiband dipole with coax and an antenna tuner (A), the SWR is low between the transmitter and the tuner and high between the tuner and the antenna (which can produce catastrophic power losses). A better arrangement would be to put the antenna tuner or antenna coupler at the feed point (B). That way, the SWR is low from the transmitter to the tuner, resulting in minimal losses. Because there is no coax between the tuner and the antenna, there can be no additional losses because of high SWR.

To effectively use coax in this case, we need to reduce the length of the run between the tuner and the antenna, and lengthen the leg between the tuner and the transmitter. For the best results, we need to put the antenna tuner at the feed point of the antenna! Remember: The rig-to-tuner leg has a low SWR! Low SWR means minimal cable losses. (See Figure 4.11B.)

Unless your arms are unusually long, working the tuner controls from 75 feet away could be a problem. What we need at the feed point is an automatic tuner, or autotuner. Autotuners that cover a range of frequencies and power levels are available from a variety of manufacturers, some even in kit form. Feeding a multiband antenna through a coax-fed autotuner can be effective *and* convenient.

TUNERS

With any sure-fire solution there's bound to be a catch or two—and antenna tuners are no exception. In the previous discussion of tuners and tuner placement we ignored tuner efficiency, tuner loss and the special transformers that are usually required to use balanced feed lines with most antenna tuners.

Like any transformer, antenna tuners are lossy. Some are more lossy than others, but no antenna tuner is 100% efficient. Loss typically increases as antenna impedances move up or down from 50 Ω. As a rule, matching an antenna that's pretty close to being resonant results in minimal tuner losses, while matching largely non-resonant loads results in greater losses.

Although we can't escape tuner losses, we can take steps to minimize them. One way is to avoid using antennas at frequencies where they'll present difficult or extreme impedances. Don't try to load a 40-meter dipole on 160 or 80 meters, for example. Cut the antenna for resonance at the *lowest frequency* band of operation and use the tuner to match the loads at higher frequencies.

You can load antennas that are physically smaller, of course, but even if your tuner can match the load, efficiency is reduced, sometimes dramatically. This is also a problem for mobile operators, but we've all heard mobile stations on the air, so we know that even teeny antennas can radiate useful signals. If you can't put up a reasonably sized antenna, put up whatever you can and feed it with open-wire line or place the antenna tuner at the antenna feed point. Whatever you do, *don't* feed a physically small antenna with a long run of coax and a shack-mounted antenna tuner. Even if your tuner matches the load, you *won't* radiate much of a signal!

Most modern antenna tuners are designed to match unbalanced loads that are fed with 50-Ω coax. To feed antennas with open-wire line or twin-lead, a balanced-to-unbalanced transformer is required at the tuner output. These devices, called *baluns*, are widely misunderstood and often perform poorly.

Baluns, whether external or built-in, often don't work well over the wide frequency ranges demanded by tuner users. It's easy to build a high-performance single-band balun, but it's difficult (perhaps impossible) to build a good-performing wideband balun.

Air-core balun transformers are sometimes used for single-band applications, while ferrite-core baluns are used to cover wider frequency ranges. If a balun core saturates—exceeding its power or voltage-handling capacity—the performance of the transformer degrades or disappears. When this happens, the balanced feeders become unbalanced, and the feed line can lose its isolation from the load (it can become part of the antenna, radiating RF energy and potentially causing RFI and other problems). Remember: Feed lines shouldn't radiate—that's what antennas are for!

The design of a balun can also contribute to its balance (or lack thereof). Typical antenna tuners use inexpensive voltage baluns, which attempt to equalize the RF voltages present in each leg of a balanced feed line. Current baluns, preferred by many experienced antenna experimenters, attempt to balance the RF current flowing through each leg of a balanced feed line. From my perspective as an end user, the experts seem to prefer current baluns. Your mileage may vary!

Balanced feed lines should remain balanced, and anything that unbalanced them is generally undesirable. I'm speaking from experience! To help ensure good balun performance, consider ignoring the small built-in voltage baluns found in most antenna tuners. Instead use an external, heavy-duty current balun, as shown in **Figure 4.12**. A balun rated for 3 kW works just fine at 1, 5 or 100 W!

The most elegant way to feed balanced transmission lines is with a balanced tuner (now that's logic even I can grasp!).

Antenna Systems and Components   4-21

Figure 4.12—This MFJ high-power current balun is designed to make it easy to feed an antenna with open-wire line. The beefy balun mounts outside, allowing users to run ladder line to the antenna and coax through the wall to an in-shack antenna tuner. (author photo)

Figure 4.13—This homebrew coaxial-cable balun covers a wide range of frequencies with ease. It's designed to be used at the input of a balanced antenna tuner. (author photo)

Balanced antenna tuners haven't been in vogue commercially for quite a while, but flea market treasures can be found, and a good-performing balanced tuner is relatively easy to build. In these designs, the balun (made from coaxial cable wound onto a piece of PVC tubing as shown in **Figure 4.13**) is placed at the tuner's 50-ohm input, where it performs well over a wide range of frequencies. The balance these tuners exhibit is superior to that of conventional units that place the balun at the tuner output.

The automatic tuners found in high-end HF transceivers are simply computerized standard tuners. The autotuners we discussed earlier are more correctly called antenna couplers. As shown in **Figure 4.14**, they're usually designed to be placed at antenna feed points and use wide-range, unbalanced matching networks to do their magic. Because of this, they don't need baluns and they don't need to worry about feed line balance. With the antenna connected directly to the coupler, there *is no feed line* between the tuner and the antenna! The run of 50-Ω coax between the transmitter and the coupler is shielded, unbalanced, and enjoys low SWR and low RF power losses. As with any transformer, the autotuner's matching network isn't lossless (especially when matching very high- or low-impedance antennas), but there are no balun losses to consider, either.

Figure 4.14—This SGC-231 autotuner mounts at the feed point of a dipole, loop, vertical, whip or end-fed wire. Its wide-range matching network is activated when the unit senses RF from your transmitter. It also remembers tuning solutions for the frequencies you habitually use, and stores them for lightning fast future tune-ups. If you can figure out a way to camouflage the box (if necessary), autotuners, also known as auto-couplers, make Stealth antenna experimentation and operation a breeze. (author photo)

Antenna Systems and Components   4-23

WIRE, INSULATORS AND SUPPORT LINES

Building Stealth antennas is a lot like building conventional antennas. Some of the materials are chosen for their invisibility factor, but the electrical characteristics of each part are the same. With that in mind, let's look at a few common antenna parts and their Stealthy alternatives.

WIRE

On one hand, wire is wire. Copper is best, insulation doesn't matter unless the wire will touch trees or buildings, and stranded doesn't work better than solid (but it's better at bending, flexing and swaying in the breeze). On the other hand, as we learned before, bigger (thicker) is better. Building Stealth antennas with heavy-gauge wire is rarely practical, however, so do it if you can, but don't worry about it if you can't. If only to leer at the unbelievers among us, I've made hundreds of QRP DX contacts using antennas made from thin-gauge steel picture-hanging wire. Use copper if you can, but use whatever works if necessary.

INSULATORS

We're all familiar with typical plastic and ceramic antenna insulators. Conventional models have worked well for decades. The problem is their visibility—they can stick out like sore

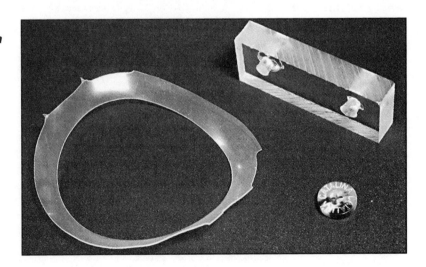

Figure 4.15—Here are several antenna insulators—from garden variety to ultra-Stealthy. (author photo)

Figure 4.16—Designed for fishing or weed trimming, modern monofilament lines make strong, invisible Stealth antenna supports. (author photo)

thumbs. When using low power (as Stealth operators should), don't be afraid to get creative with center and end insulators. Teeny pieces of clear Plexiglas work well and, depending on power levels and antenna sizes, so do pearl-colored plastic buttons and the plastic rings that keep six-packs constrained (**Figure 4.15**). I read about one QRP operator who uses a hair dryer to heat and bend plastic dinner forks to serve as twin-lead stand-off insulators. When necessary, use materials that actually *are* insulators and be creative.

SUPPORT LINES

Conventional antennas can be raised with lengths of poly or nylon rope, which are probably too visible for many Stealth installations. Coming to our rescue is heavy-gauge monofilament fishing line, or the even heavier monofilament line used in weed trimmers (**Figure 4.16**). Colorless weed-trimmer line is invisible once airborne, it's inexpensive and available at most department stores or hardware stores. To throw a line over a tree (or whatever) I usually start with fishing line which, once in place, I use to pull up the weed trimmer line. These heavy monofilament lines have survived several Minnesota winters (and summers) without breaking.

NOTES
[1]Ford, Steve, WB8IMY, "The Lure of the Ladder Line," *QST*, December 1993, pp 70, 71.

Chapter 5

Stealth Antennas

Now that we've reviewed several concepts and components that are important to Stealth installations, let's look at a bunch of antennas. As you can imagine, there are hundreds of Stealth antenna possibilities, but the most effective and practical designs tend to be variations on a few basic themes. There's no magic here. Just as "regular" dipoles and loops work well, so do their Stealth counterparts.

So what *are* the big differences encountered when building Stealth antennas? The most prominent are concealment and disguise. A disguised vertical is still a vertical, an attic dipole is still a dipole, and so on. Electrically, Stealth antennas have no idea that they're "special."

Unlike regular antennas, however, Stealth antennas are often erected in close proximity to other objects—houses, trees, attics, shingles and gutters—or they're shaped a bit differently to accommodate available spaces. These factors *can* reduce an antenna's effectiveness and change its resonant frequency. For example, if you tune a 40-meter dipole for use in the clear, when you string it under the eaves of your house (bending a few feet of each end along the side of the house, perhaps), the resonant frequency will be slightly different.

You'll have to "trim" the antenna once it's installed. You

may also discover that the lowest SWR you'll be able to attain is also different (and probably higher). The "de-tuning" effects of nearby objects can usually be corrected by re-tuning or by using an antenna coupler or tuner. Because most Stealth operators prefer to be active on several bands, a coupler or antenna tuner will probably be required anyway.

Before we get started, let me again remind you that I'm assuming you have a basic knowledge of how to build wire Amateur Radio antennas, how to attach center and end insulators, how to solder coax or open-wire line to an antenna's feed point, and so on. If you're new to this aspect of Amateur Radio, check out a few of the helpful references in Chapter 4. They contain step-by-step instructions on assembling wire antennas, soldering coax connectors and installing insulators. Also, if necessary, try to enlist the help of a more experienced antenna builder to help you get started.

To present as many Stealth antenna ideas as possible in this chapter, I didn't provide precise construction details for every antenna. Most of the "common knowledge" details have been left to your discretion. Every installation is unique, but, after all, rigging one center insulator is like rigging another; soldering one coaxial cable is pretty much like soldering another, and so on.

LOOPS, DIPOLES AND Vs

Balanced antennas—loops, dipoles and Vs, among others—have been popular and effective choices for millions of ham, military and commercial operators for decades. They're easy to build, forgiving in nature and offer typical Stealth operators a variety of benefits. Because these "self-contained" antennas do not need RF ground connections, they may work better in compromising situations.

You're probably familiar with the shapes of dipoles and loops, and with the formulas used to calculate the element lengths (468 divided by the frequency in megahertz for dipoles; 1005 divided by the frequency in megahertz for full-wave loops). The feed point impedance of each provides a reasonable match for standard 50-Ω coax. Each can also be fed with 450-Ω open-wire line or 300-Ω TV twin-lead and an appropriate antenna tuner.

Common dipole installations include the flat-top (**Figure 5.1A**); the inverted V, which requires a single elevated

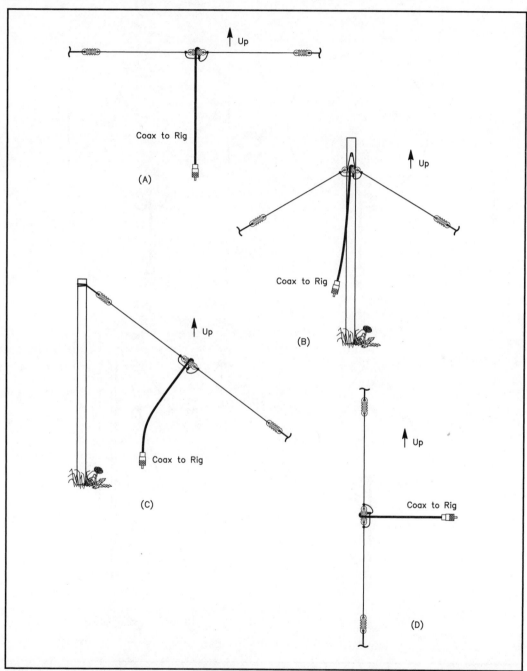

Figure 5.1–Standard dipole configurations: (A) flat-top, (B) inverted V, (C) sloper or sloping dipole, and (D) vertical dipole.

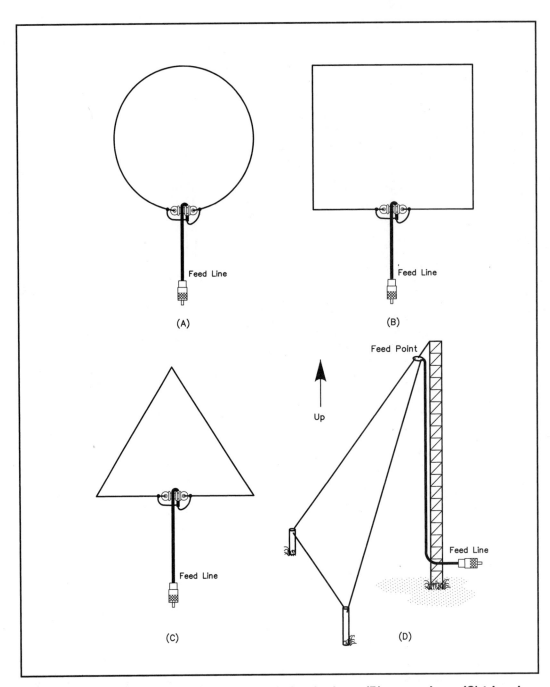

Figure 5.2–Standard loop configurations: (A) circular loop, (B) square loop, (C) triangle or delta loop, and (D) sloping delta loop.

support (Figure 5.1B); the sloper, a dipole that's slanted at about a 45° angle (Figure 5.1C); and the vertical dipole, which runs straight up and down (Figure 5.1D). The vertical configuration is the least common. To minimize feed line interaction, feed lines should be placed at a 90° angle to the plane of the wire elements. For flat-tops and Vs this is fairly easy. For slopers and vertical dipoles it becomes more difficult.

Loop antennas are theoretically circular (**Figure 5.2A**), but the square shape of the quad (Figure 5.2B) and the triangular shape of the delta loop (Figure 5.2C) are most common. The plane of a loop can be oriented vertically, horizontally, or even at an angle (like the sloping delta loop shown in Figure 5.2D).

In most non-Stealth installations these shapes can be preserved. In many Stealth situations, however, we need to get creative to fit the antennas to whatever physical space is available. Thanks to the accommodating nature of these designs, we can do just that.

So, when is a dipole a dipole and a loop a loop? From an antenna perspective you'd be surprised. Take a look at the shapes shown in **Figure 5.3**. These are all dipoles, Vs and loops and, for the most part, will perform and tune up accordingly.

Let's say you're stringing a dipole under the eaves of your house. If the leg lengths of your 40-meter dipole are 10 feet too long, rather than making the dipole shorter (increasing its resonant frequency and reducing its performance on the lower bands), simply bend the ends around the edges of the house. In general, you should try to keep about half of each dipole leg on the straight and narrow. Beyond that, the ends can be bent back, forth, up, down, sideways or whatever. The only thing you can't do is fold the legs back along themselves. That's almost like cutting the elements with a wire cutter! For loops, try to maximize the enclosed area while keeping the wire segments as far apart as possible. Configurations to be avoided are shown in **Figure 5.4**.

The same flexibility is afforded to loops. While writing this book I built a Stealth loop that followed the underside edge of the overhang of my garage roof. I used six-inch strips of Plexiglas attached to small metal "L" brackets as stand-off insulators. On two sides, the plane of the loop was perpendicular to the ground, but on the gable ends, the wire followed the roof line and "peaked"

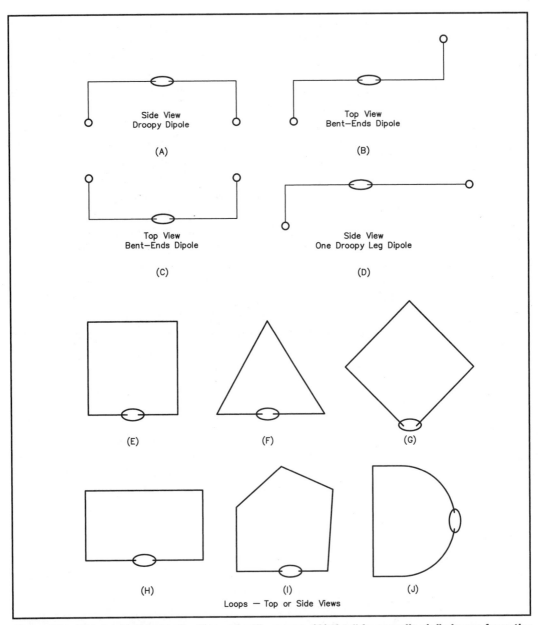

Figure 5.3–Dipoles and loops for "imperfect" spaces: (A) the "droopy dipole" shown from the side, (B) the zig-zag dipole viewed from the top, (C) the "bent-end" dipole (often used under the eaves of a house) viewed from above, (D) the "one-legged droopy dipole" viewed from the side, (E) square loop, (F) delta-shaped loop, (G) diamond-shaped loop, (H) rectangular loop, (I) "random" or ugly loop, (J) "Big D" loop. Items (E) through (J) can be top or side views, with the feed point placed anywhere.

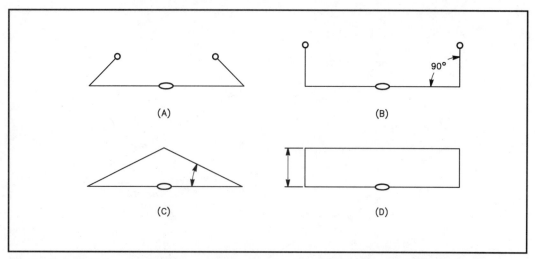

Figure 5.4–Feel free to adapt common loop, dipole and V designs to fit your particular space, but don't fold the legs of a dipole back on themselves as shown at A—try to keep the droopy or bent ends to 90° or more as shown in B. Also don't run the wires of triangular or rectangular loops too close together, like the shapes shown at C and D. Try to enclose as much area as possible!

at the roof ridge. The garage is roughly 25 feet on a side. Accounting for the extra wire for the runs on the two gable ends, and for the size of the overhang, the loop had about 28 feet of wire on each leg.

The shape of the loop is shown in **Figure 5.5**. Although it was only eight to 15 feet above the ground, this antenna tuned and worked well on all bands from 160 through 6 meters. I fed it with a six-foot length of heavy-duty 50-Ω coax. While making many contacts over several months, I found that the Stealth loop was usually one to three S-units below the performance of my main antenna, a 5.8-MHz horizontal loop that's 40 feet above the ground and fed with 450-Ω ladder line. For a Stealth antenna that covers all bands and is easy to tune, however, it's a fine antenna indeed. If my roof were on the second, third or fourth floor, performance would have been nearly indistinguishable from a traditional loop. When I lived in Connecticut, I was fortunate enough to have space for a 40-meter horizontal loop inside my fourth-floor "walk-up" attic. I had used outdoor horizontal loops

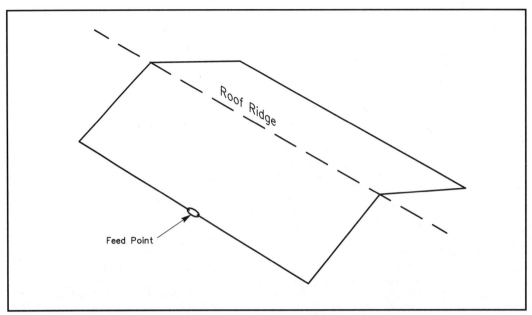

Figure 5.5–This under-the-eaves loop, which encircles the author's garage, contains a little more than 100 feet of wire. Although it's only 8 to 15 feet above the ground, it's a good, Stealthy performer on all bands.

for years, and when using the attic loop on the air I couldn't tell that the antenna was mounted inside.

Now that you're ready to be creative with the shapes of your dipoles, loops and Vs, let's look at some practical designs.

BACK LOT SPECIALS

If your lot contains woods, swamps or other areas that people tend to avoid, you may be able to erect full-size conventional antennas—even small towers and beams—that are hidden in the "back forty." In these cases your feed line is probably the prime concern. You'll want to identify ways to hide the part that might be visible. Long coax runs are okay for matched lower-band antennas and may be okay for antennas that have feed-point-mounted autotuners. For multiband operation with your shack-mounted tuner you'll need to use 450-Ω ladder line, which can be run for several hundred feet under the right conditions.

For example, let's say we want to put up a 40-meter dipole in

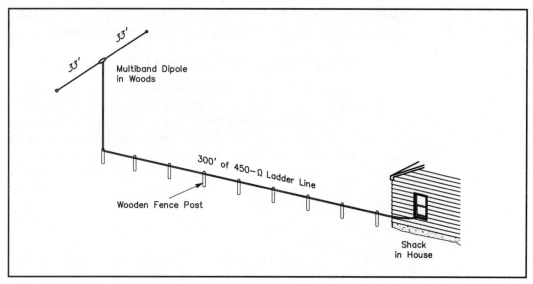

Figure 5.6–Under certain conditions, 450-Ω ladder line can be practically lossless, even over distances of several hundred feet! See text.

the woods behind our duplex. We don't have an autotuner so we'll have to tune the antenna on all bands from 40 meters and higher frequencies. To ensure that the antenna isn't discovered, let's place it a whopping 300 feet from the shack. That's a heck of a run for coax — even good-quality coax! (See **Figure 5.6**.)

At the design frequency of 7 MHz, though, coax losses are only a little more than 1 dB. On 10 meters, however, we'll lose 33 dB between the transmitter and the antenna—*almost all* of our power would be wasted! On 80 meters, a band that's below our design frequency (a problem for long coax runs), a 300-foot run of coax would consume more than 80 dB of our transmitter power. We'd be better off using a ground-mounted mobile whip!

With the ladder line, however, things are pretty rosy. On 40 through 10 meters, feed line losses for a 300-foot run average about 0.6 dB! On 80 meters the loss would be about 8 dB. That's not fabulous, but it's a lot better than the 80 dB we'd lose with coax. To improve low-band performance we could put up a center-fed 80-meter dipole (doubling the size of the hidden antenna).

• Advantages: These antennas are truly hidden. If necessary, you can use greater output power.

- Disadvantages: Long cable runs can be a pain—especially ladder line, which can't be buried or run at ground level.

INVISIBLE DIPOLES AND LOOPS

Because outdoor antennas tend to outperform their indoor counterparts, you should consider erecting an outdoor dipole, V or loop, if possible. If your house has a reasonably shielded or secluded backyard, an "invisible" dipole or loop made from 28-gauge enamel wire, small Plexiglas insulators and heavy-duty monofilament support lines might be perfect. When viewed from the street, these antennas are truly invisible. Remember to keep the feed line out of sight and "blocked" by the front of the house. You'll probably have to support the feed line, which weighs more than the antenna!
- Advantages: Outdoor antenna performance with "no visible means of support."
- Disadvantages: Antennas made from fine wires tend to break more often than those made from beefier materials. You may have to put them back up every now and then.

HOUSE-MOUNTED DIPOLES AND LOOPS

If you don't have a secluded backyard, and traditional invisible antennas might invite scrutiny, perhaps you can mount your dipole or loop on your house or apartment. Mounting dipoles and loops under the eaves is a perennial favorite. This works especially well with wood-frame houses that are sided with wood or vinyl. If you have steel or aluminum siding, choose another Stealth antenna!

When installing "under the eave" dipoles and loops: use short stand-off insulators, if possible. Make the loop or dipole as large as possible (which helps low-band performance) while still keeping it hidden; feed the antenna with open-wire line, if possible (coax is okay if the cable run is short); and feed dipoles at the center (loops can be fed anywhere). Remember: The antenna wires can be shaped (within reason) to fit your space.

In my experience, eave-mounted loops usually work better than dipoles. Your exact requirements, however, will dictate the type you should install. By the way, the Christmas season is a good time to string these types of antennas. If the neighborhood committee gets nosy, you're simply putting up a "stringer" wire for your new Christmas lights!

If your need for secrecy is extreme, consider adding plastic rain gutters to your house. You can string a Stealth dipole inside the outer edge of the protective gutter. Make sure the dipole is suspended in the center of the rain gutter or attached to the outer edge and not lying on the bottom of the gutter where it might collect moisture and debris.

- Advantages: These antennas are easy to erect and camouflage, and they work better than most indoor antennas.
- Disadvantages: Because the antenna elements are often in close proximity to living areas and house wiring, RFI and received noise can be a problem. Low-power operation is mandatory.

MOBILE WHIP DIPOLES

Although we're used to seeing mobile whips mounted on cars (and occasionally on balcony railings), a pair of mobile whips mounted back to back can create a usable low-profile dipole antenna that's suitable for backyards, balconies or "quick-setup" portable use.

Because the whips are resonant at a chosen operating frequency, they "take power" without a tuner, and load up well. You can make an antenna of this type by making a dipole from two identical mobile antennas that are resonant at the same frequency. (See **Figure 5.7**.)

Some mobile antennas, such as the *Spider*, feature multiple resonators mounted on a single mast. Mounting two of these back to

Figure 5.7–Take two identical mobile whips, mount them back to back and connect the coaxial center conductor to one whip and the shield braid to the other. You have an instant resonant dipole! See text.

back can provide a coax-fed multiband dipole (40, 20, 15 and 10 meters) that's lightweight and measures only about 12 feet from tip to tip.

Mount the resulting dipole horizontally or vertically, but keep it at least a few feet away from structures, people, and other conductive objects. Some operators even turn these antennas with small TV rotators. On 10 and 15 meters, where the 6-foot whips are closer to full-size, you might even notice some directivity.

- Advantages: Mobile whip dipoles are easy to make and quick to set up and take down.
- Disadvantages: Physically short antennas have narrow SWR bandwidths. On 40, 20 and 15 meters you'll have to tune the whips to cover your favorite—and perhaps quite narrow—parts of the bands. On 10 meters, however, useful bandwidths are easier to achieve.

INDOOR LOOPS AND DIPOLES

If outdoor antennas of any type are out of the question, indoor antennas will have to do. If you live in a wood-frame building (or almost any non-steel structure), indoor antennas will get you on the air, but your signals will probably be a bit weaker and you will have to deal with the noise, RFI and RF exposure factors that accompany them. Again, low-power operation is a must. Placing medium- and high-power transmitting antennas in living areas is not acceptable! See the RF Exposure sidebar in Chapter 1.

Installing indoor loops and dipoles is mostly a matter of aesthetics and negotiation—other people who share your living space may not want an inverted V antenna suspended from the cathedral ceiling! Dipoles and loops can be tacked, taped or otherwise fastened to walls and ceilings. The edge where the walls meet the ceiling is a time-honored favorite. Full-size antennas for 30 meters and longer wavelengths represent an additional challenge for most indoor locations. There's probably not enough room! If you have to use smaller antennas to fit available spaces, your antenna tuner or coupler will probably match the antenna, but the overall efficiency will suffer with smaller antennas.

You can make a functional multiband indoor antenna by running an insulated, heavy-gauge wire around the perimeter of the ceiling. Don't try to cut the loop to any specific frequency; simply cut the antenna to fit the available space. Feed the loop with open wire line (preferred) or coax (if the run is quite short or if the antenna is in the same room as your transmitter) and tune it with your antenna tuner or autotuner/coupler. Think second or third floor—even the attic—if possible.

As an alternative, you could place the loop wires around the perimeter of one wall—a vertically oriented loop. Tuning and feeding considerations are the same as for the ceiling loop. Dipoles can be installed and tuned in a similar fashion, but when you can put up a loop with no additional effort, the extra wire will almost always improve the performance in indoor settings.

If wiring up a wall or the ceiling is too intrusive and you have a wide-range tuner or autocoupler you can "scramble wind" 70 to 90 feet of insulated wire into a square, wall-mounted loop that is at least two to three feet on a side. (See **Figure 5.8**.) The antenna shown here is fed with an autocoupler. It's better than no antenna,

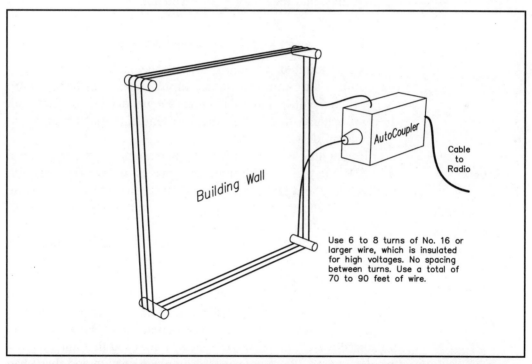

Figure 5.8–For a temporary or "last-ditch" indoor antenna, scramble-wind 70 to 90 feet of insulated wire into a single loop (about three feet on each side). For ease of use, place an auto-coupler at the feed point and feed the antenna with coax. This arrangement may work with a conventional tuner. If you use one, place the tuner at the feed point–don't run coax from the tuner to the antenna! (Antenna courtesy of SGC's Smartuners for Stealth Antennas, www.sgcworld.com).

of course, but its performance will probably not match that of a larger indoor loop. Such a small loop may also have sharp nulls and may even be highly directional.

I've occasionally daydreamed about making a small scramble-wound loop on a wooden frame and turning the loop with an "upside down" ceiling-mounted antenna rotator (or a floor-mounted swivel for "armstrong" rotation). Being able to turn the loop may allow for signal peaking or noise nulling.

In keeping with an earlier promise, the loops described here are simple and easy to build. Small loop antennas comprise their own esoteric field of study. For information on more sophisticated designs, see the references at the end of this chapter.

- Advantages: Indoor loops are cheap, quick and easy to erect.
- Disadvantages: They are prone to noise pickup and RFI problems, and they may expose you and your family to excess RF energy.

ANTENNAS IN THE ATTIC

Attic-mounted antennas are a special case. With few exceptions, attic-mounted loops, dipoles and mini beams work better than the other indoor antennas we just discussed. That said, it's unfortunate that some attics present potential obstacles to antenna builders.

If your attic is laced with power and telephone wires, jammed with RF-noisy air conditioners or furnaces, or overflowing with stored household junk, installing attic antennas can be a problem. If your highest household space is relatively uncluttered, however, you're in business.

Earlier I mentioned my multiband attic loop and praised its performance. That antenna lived in a spacious fourth-floor "walk-up" attic that seems to be popular in some parts of the country. The loop ran around the perimeter of the floor. On the east end of the attic I strung a two-element tri-band delta loop beam between the attic's high ridge pole and the wooden floor! On the west end I had a three-element 2-meter beam. I attached that antenna to a mast-mounted TV rotator on a large piece of plywood that I had put on the floor. I later replaced the 2-meter beam with a commercially made HF mini beam.

My former Connecticut QTH represents attics at their best. In the Midwest, where I now live, most attics have been finished off as living spaces (one and a half story houses are predominant) or are mostly inaccessible (too "short" to move around in). If your attic is big and beautiful, go ahead and fill it with antennas!

The best all-around attic antenna is probably a venerable horizontal loop fed with coax or ladder line and trimmed with an antenna tuner or an autocoupler. Don't worry about cutting the loop to a certain frequency. Simply run it around the perimeter and feed it wherever it's handy. Use stand-off insulators to keep the wire suspended in space.

If you have a hidden or protected backyard and a suitably oriented attic, you might consider making a "half in, half out" multiband dipole. As shown in **Figure 5.9**, one leg of the dipole is in the attic, while the other leg is run to a tree or other suitable support in the backyard. The antenna is fed with 450-Ω ladder line, which drops away vertically to the shack window below. To take advantage of such an installation, you have to have the right building and a workable shack location, but if everything's in place, the "HIHO Attic Special" can help you put up a large dipole that offers improved performance on all bands, especially the lower-frequency bands.

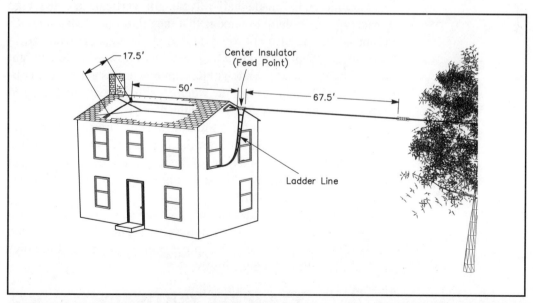

Figure 5.9–If you have a suitably situated attic, because of its size, the half-in, half-out dipole can help you put out a big signal, even on the low-frequency bands. See text.

Large attics or empty upstairs bedrooms are perfect for housing VHF/UHF beams. They might also accommodate compact HF beams. If you have the space, use your imagination. My small 6-meter beam is mounted on an 18-foot-high conduit mast bolted to the edge of the garage roof. If I could mount it in the attic it would be 10 feet higher and might even work better. Because that space is a teenager's bedroom, however, I think I'll leave it on the mast!

- Advantages: Attics are great if you have them...
- Disadvantages: ...and nonexistent if you don't. Attics can also be filled with RF-noisy motors and machinery.

VERTICALS AND INVERTED L ANTENNAS

Most hams either love vertical antennas, or they hate them. Their performance often depends on the quality of the RF ground over which they are installed. The same goes for inverted L antennas, which are really "bent" verticals in disguise. Without a decent RF ground or a good counterpoise, they don't put out much of a signal. On the other hand, when properly constructed, these antennas can easily be hidden and can work quite well, often on several bands.

Commercially made half-wavelength verticals are the rage nowadays. Their claim to fame is that they don't need elaborate RF grounds. They're also not too Stealthy, although some hams go so far as to put them up at night and take them down during daylight hours! The "Cadillac" version of this approach uses a motorized tip-over mount. The antenna lies flat when Stealth is needed and is raised to its regular vertical position when the coast is clear!

If a situation called for a Stealth antenna "crash dive," I can imagine myself running through the house yelling, "Dive, dive, dive!" like the captain of a beleaguered submarine. "Let's get that antenna down, people. Our standing with the Neighborhood Association depends on it!"

If you don't have the resources (or the sense of humor) for a tip-down Stealth mount, many half-wave verticals can be hidden inside thin-wall plastic or PVC pipes and disguised as flagpoles or Space Needle birdhouse towers.

FLAGPOLE VERTICALS

Vertical antennas disguised as flagpoles seem especially

popular in the southern coastal states, where the soil has excellent conductivity (good RF grounding) and the laws seem to guarantee a citizen's right to erect a flagpole—no matter what deed restrictions may be in effect.

In addition to hiding commercially made half-wave verticals inside plastic or PVC pipes, typical designs run quarter-wave wire elements inside the plastic pipes and require the use of one or more quarter-wavelength radials (counterpoises) that are typically buried in the lawn. If you put up such an antenna, rent a hand-operated lawn trencher to conveniently cut a shallow slit in the grass. Once the wire is buried, the slit will quickly heal. If you try to use a shovel or some other ill-advised tool, who knows what might happen!

NOTE: Whenever working with verticals, inverted Ls or end-fed wires, make sure your radial or counterpoise wires are physically or electrically longer than the radiating element (so your antenna tuner knows which is the "ground" and which is the "antenna").

Version 1

The simplest flagpole verticals are made from two lengths of PCV pipe (telescoped, perhaps) cut to the required height. Hidden inside is one or more quarter-wavelength insulated wires. It's easy to make a multiband vertical that covers the ham bands from 20 through 10 meters. Simply cut five quarter-wave wires, one each for 20, 17, 15, 12 and 10 meters, connect the wires at the base (feed point) of the vertical and run the other ends up into the PVC pipes. Cut five matching quarter-wavelength wires for the counterpoises, which will be buried in the lawn (the five counterpoise wires are connected at the feed point). Use a small waterproof box at the base to hide the connections. The design is detailed in **Figure 5.10**.

You'll probably want to paint the PVC with non-metallic silver paint and glue a plastic ball to the top of the pole with silicone sealer (to keep out water and to keep up the disguise). Instead of using a single quarter-wavelength ground radial on each band, some operators run a bunch of ground radials in several directions (make sure the radials are at least a quarter-wavelength at the lowest operating frequency).

Because the antenna is resonant (or nearly so) on several

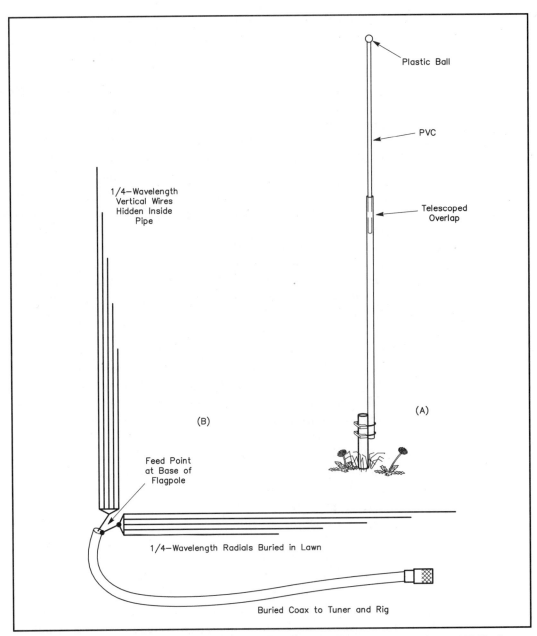

Figure 5.10–(A) One or more quarter-wavelength wires hidden inside a length of PVC pipe make an excellent Stealth vertical antenna. (B) Matching radials can be buried in the lawn. Glue a plastic ball to the top; paint the pipe and ball with non-metallic gray paint; add minimal rigging for a flag (only if necessary!); camouflage the feed point according to your needs; and bury the coax and radials in the lawn. The result is a classic flagpole vertical. See text.

bands, excessive SWR losses in the buried coaxial feed line will probably not be a problem. Make sure your coaxial cable is rated for direct burial or it may quickly become contaminated. If you can't obtain direct-burial coax, you might get away with burying regular coax and replacing it every year. Alternatively, you might experiment with burying the coax inside a suitable length of garden hose.

Because the wires that make up the vertical and counterpoise elements will interact, pruning and tuning for the best SWR on each band can be time consuming and may draw suspicious looks from neighbors, who are expecting to see a simple "flagpole raising." Because the SWR losses of these "reasonably resonant" wire verticals will probably be minimal, use your shack-mounted tuner to tweak things. For a luxury installation, place an autotuner or autocoupler at the base of the flagpole. A 20-meter vertical is just less than 17 feet tall, which is probably a practical upper limit for flagpole verticals made from PVC tubing.

Version 1 Up a Tree

If you have a tree that's at least 17 feet tall in your backyard, an interesting variation of the Five-Band flagpole vertical can be made by running the insulated wires up the tree trunk (with the feed point and radial wires connecting at the base of the tree). This antenna is shown in **Figure 5.11**. Will the tree absorb some of the signal? Yep. It will work just fine, however, and it will save you the hassle of erecting a flagpole!

One advantage to the tree design is the possibility of working 30, 40 and maybe even 80 meters. You can easily throw a piece of monofilament line over the top of a tall tree and use it to pull up a length of "hard to see" enamel-insulated wire.

Version 2

If you have to put up an honest-to-goodness aluminum flagpole you can insulate the base from ground by mounting the pole on an insulated post (or inside an insulated sleeve) and working the "vertical" flagpole against a few radial wires buried in the lawn. This is essentially a large mobile whip worked against a radial system instead of a car body. Because the flagpole may not be resonant on any ham bands, using a shack-mounted antenna tuner may invite excessive feed line losses. These antennas

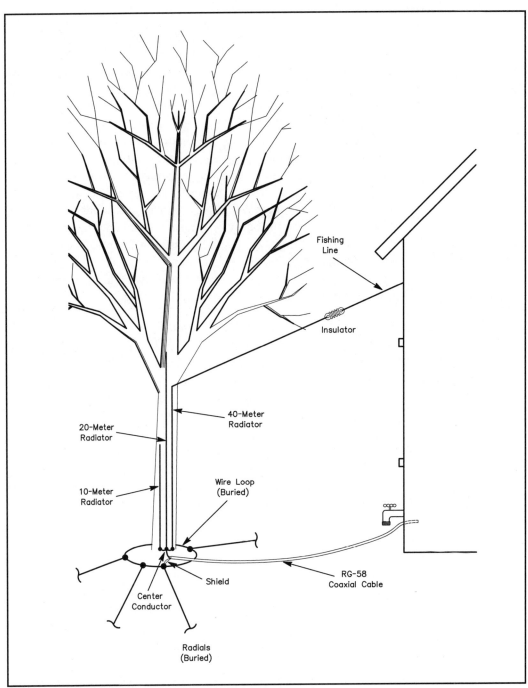

Figure 5.11–If you have a handy tree, you can dispense with the flagpole! See text.

usually work best if you use an impedance-matching transformer (single band) or an autocoupler at the feed point.

If your existing flagpole is already grounded it may still be possible to shunt feed the pole on several bands. That's beyond the scope of this book, however, so you'll have to seek a solution in the *ARRL Antenna Book*.
- Advantages: Flagpole and tree verticals turn "neighbor-approved" objects into hidden HF antennas!
- Disadvantages: You may be band-limited depending on the size of your flagpole. Sixty-six foot flagpoles work well on 80 meters, but your neighbors will likely be more than excited!

INVERTED L ANTENNAS

The inverted L antenna is a "bent-over" vertical antenna in disguise. Because you can make them from fine insulated wire,

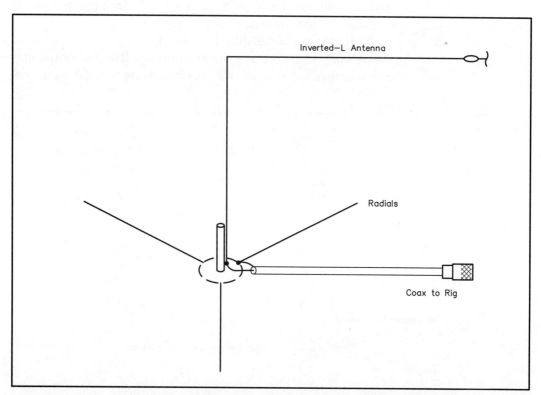

Figure 5.12–This diagram shows an inverted L antenna with three radials.

however, they're often easy to conceal in backyard trees or against the side of a house or building.

The basic inverted L is shown in **Figure 5.12**. Like any other vertical, it requires a counterpoise or radial system. Unlike flagpole verticals, which tend to be size-limited, inverted Ls can be rather large, which makes for decent performance on the low-frequency bands. Although an inverted L can be cut for a specific band ($1/4$ wavelength or $3/8$ wavelength, for example), the rule for Stealth L antennas is to make them as large as space allows.

If you have two appropriately positioned trees in your backyard, throw a monofilament line over the top of each and use them to pull up a thin, enamel-covered wire. Install a similarly sized counterpoise or radial system (buried in the lawn or run around the foundation of the building). Feed the inverted L with coaxial cable (buried or lying on top of the grass). This excellent performer is shown in **Figure 5.13**. Because the antenna is probably not resonant, a shack-mounted antenna tuner may be lossy. A feed-point-mounted autocoupler is ideal for multiband operation.

A similar super-Stealthy Inverted L can be fastened to the side of your house or apartment building. Run the horizontal portion across any non-metallic roof or string it under the eaves

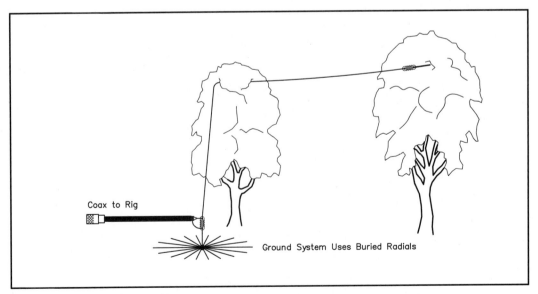

Figure 5.13–Two tall trees can support one large Stealth inverted L antenna.

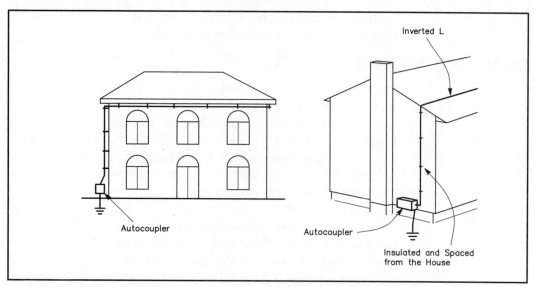

Figure 5.14—The inverted L antenna can be very Stealthy. If your house or apartment has two or more floors, hide the horizontal run under the eaves or on top of the roof (wood-frame building/roofs only).

using suitable stand-off insulators. See **Figure 5.14**.
- Advantages: Large inverted L antennas can work quite well on the low bands. Most versions are easy to conceal.
- Disadvantages: Like any vertical, inverted L antennas need a counterpoise or radial system to be effective. Best performance is often achieved with feed-point mounted autocouplers.

MOBILE WHIPS ON THE BALCONY

If your apartment, condo or hotel room is on the 40th floor and sports a sizable metal balcony railing or fire escape, you may be able to use a conventional mobile whip worked against the railing. This tends to work best at VHF or on the higher-frequency HF bands, where the railing or fire escape is a "larger" RF ground. On the lower-frequency bands the setup may not be too efficient, especially when you consider that a high-quality 80-meter mobile whip is about 6 to 8% efficient on a good day! Feel free to enhance your performance by using an appropriate counterpoise (run around the baseboards, perhaps).
- Advantages: Cheap and easy if your balcony is a good counterpoise.
- Disadvantages: They're poor performers if it's not. Mobile

whips, no matter where they're mounted, have narrow SWR bandwidths and inherent inefficiencies.

END-FED WIRES

For some reason—probably *not* their performance—end-fed wires are synonymous with Stealth operation. At their romantic best, end-fed wires appear to be a Stealth operator's cure-all. Simply string a random-length wire between a backyard window and a handy tree, feed it with an antenna tuner that's grounded to a nearby water pipe and you have an instant antenna — or an instant headache!

End-fed wires *need* counterpoises or other suitable RF grounds to be effective—and often to keep unwanted RF out of the shack and elsewhere. Although others occasionally laud the performance and convenience of end-fed wires, I must admit that I haven't had a lot of luck with them, with one exception.

When I moved to Connecticut I lived in a hotel room for three months while I looked around for an apartment. To get on the air I ran a 100-foot length of 28-gauge steel wire from my second-story balcony to a willow tree in a wooded area near the edge of the hotel's back parking lot. I used a standard antenna tuner to feed the wire and a counterpoise wire strewn around the room as an RF ground. When driven by a borrowed Heathkit HW-9 QRP transceiver, I made easy stateside and DX QSOs on CW with no detectable RFI. This classic "random wire" design is shown in **Figure 5.15**.

If you attempt to use a similar end-fed wire antenna, make sure the wire—and the counterpoise—is at least a quarter-wavelength long at the operating frequency. This type of installation seems to benefit from the use of artificial grounds or counterpoise tuners. If you have problems with RFI or RF in the shack, give them a try.

If my former Connecticut QTH had not been endowed with a spacious attic, I might have clamped an end-fed wire to my crank-out clothesline. Because I lived on the third floor, the line was 35 feet above the ground and about 50 feet long, terminating at an empty telephone pole in the backyard. If the antenna was somehow detected, I could have simply cranked it inside the back porch/landing.

- Advantages: Cheap and dirty.
- Disadvantages: Potential performance and RFI issues.

EXISTING STRUCTURES

Loading an existing structure as an antenna is rarely a good

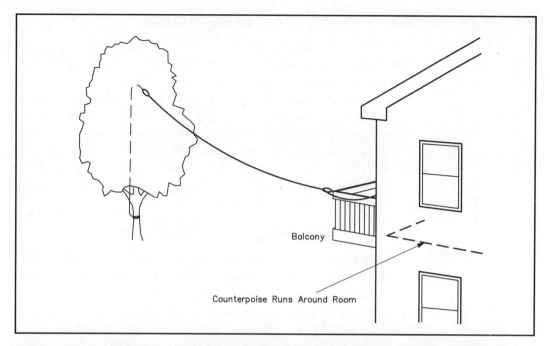

Figure 5.15—The author ran an end-fed wire from a second-floor hotel room balcony to a willow tree in the back lot. He used the balcony and a large counterpoise strewn around the room as an RF ground. The antenna worked well, even at QRP levels. Low power ensured his Stealthy status with the rest of the guests.

first choice, but it's occasionally a workable last-ditch effort to get on the air. As in my previous discussion, gutters and downspouts are often the most logical choices, especially if they're aluminum or copper. Treat them as end-fed wires and use a tuner and a good RF ground or counterpoise. You may want to securely bond the joints between sections, as "press fit" or pop riveted connections don't work very well at RF!

You can also load patio umbrellas, window frames and other existing structures. As to effectiveness—your mileage may vary! Although they can sometimes be loaded in a pinch, avoid iron and steel structures. They're probably too lossy to be effective.

- Advantages: These antennas are impossible to detect visually and provide a last-chance way to get on the air.
- Disadvantages: Antennas made from existing structures may not radiate well and users may encounter RFI problems.

COMMERCIAL STEALTH ANTENNAS

Because hidden antennas are becoming increasingly necessary, manufacturers are producing off-the-shelf Stealth antennas that may be practical at your QTH—especially if you can't erect a homemade Stealth antenna or simply want a more convenient solution. Although the commercially made antennas are handy, keep in mind that most of the homemade Stealth antennas we've just discussed will probably outperform these commercial compromise antennas. I've listed only a few here. This list of commercial Stealth antennas is far from complete.

MFJ MINI LOOP ANTENNAS

MFJ's mini loop antennas—a mainstay for travelers and condo dwellers—aren't Stealth antennas in the traditional sense. They're more properly low-profile antennas. Because they don't look like traditional antennas, however, they can often be left in plain sight. "That? Oh, that's a circular bird perch that my grandson made me in modern art class." You get the idea!

Shown in **Figure 5.16**, the Model 1786 covers 30 through 10 meters (inclusively). A small tuning box (with an SWR meter) is

Figure 5.16–MFJ's Model 1786 compact transmitting (and receiving) loop covers 10 to 30 MHz and can be mounted horizontally or vertically. Using the loop—any physically small loop—can be tedious, but the antenna's overall good performance has made it a favorite for many Stealth operators. (author photos)

mounted in the shack and a single coaxial feed line runs to the antenna, which is a small loop tuned by a motor-driven variable capacitor. Because the bandwidth of physically small loops is extremely narrow, you'll have to retune if you move a few kilohertz on either side of your present operating frequency. Although it's sometimes frustrating to tune, tune, tune, the narrow bandwidth works wonders on receive, especially if your radio needs help in the selectivity department. Off-frequency signals totally disappear!

MFJ's mini loops perform well if they're mounted in the clear. I've used them on several occasions and have been impressed with how well they work. As with all antennas, mount them as high as possible. These loops usually don't work well when mounted indoors, near the ground, or near other objects, especially metal objects.

THE BILAL ISOTRON

The Bilal Company of Florissant, Colorado, makes a complete line of Isotron antennas that covers 160 through 10 meters. These compact, tunable antennas do not need radials or antenna tuners. Although I've never used one, some hams claim they make excellent, compact Stealth antennas for 160 and 80 meters. See **Figure 5.17**.

BUTTERNUT BUTTERFLY BEAM

With a turning radius of less than seven feet, Butternut's HF5B Butterfly Beam provides a bit of gain and directivity on all bands from 20 through 10 meters in a package that's small enough to be mounted in the attic or the backyard. When

Figure 5.17–The Bilal Isotron antennas are compact single band units available for all bands from 160 through 10 meters. The 40-meter Isotron, shown here, measures approximately 21 inches high and 18 inches across.

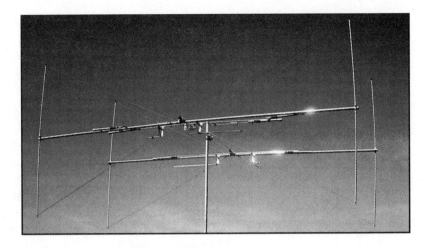

Figure 5.18–The Butternut Butterfly HF5B antenna covers the 20 through 10-meter bands. With a 6-foot boom and 12½ foot element length, this is a compact two-element antenna.

mounted in my fourth-floor attic, I was generally pleased with the Butterfly's performance. Assembly and tuning, however, aren't for the faint of heart! Setting up one of these somewhat expensive little beams definitely takes patience! Once set up, however, the mini beams perform as advertised. See **Figure 5.18**.

VENTENNA

Made by the Forbes Group of Rocklin, California, the Ventenna is an ideal solution for antenna-restricted hams who want to put up an effective VHF/UHF antenna in plain sight. Available for most ham bands from 2 meters and higher frequencies (plus a 2-30 MHz receive only SWL antenna), the Ventenna looks like a roof-mounted bathroom vent. Who in your neighborhood could object to a bathroom vent?

WHY DIDN'T I THINK OF THAT?

I can't cover every conceivable Stealth antenna in this one volume, so I hope you've learned enough to adapt one or more of the designs we've discussed here to your unique situation. I have two final solutions to offer as a salute to ham ingenuity!

Mobile Whips Mounted on the Car

If your mobile antenna works well when in motion, it'll work well when sitting in your driveway! More than a few hams who are hampered by antenna restrictions simply string a temporary run of coax from their house-mounted shacks to their driveway-mounted mobile antennas. Be sure to disconnect the "base station" coax before driving off!

The Mobile Tower and Beam Trick

I don't know if anyone has actually tried this, but the idea really intrigues me. I heard about this antenna idea while tuning the bands one day. A ham who was experiencing a lot of grief from regulation-toting neighbors and members of the Neighborhood Association was fielding advice about putting up a decent antenna from other hams on frequency. Because the ham's deed restrictions allowed no yard or house-mounted antennas of any kind, one insightful roundtable member suggested that the inquiring ham buy an old pickup truck, park it in the driveway and mount a 30-foot fold-over, crank-up tower and beam in the box. As long as the truck is licensed, insured and able to drive away under its own power (with antennas retracted), it would likely satisfy the antenna rules to the letter. Imagine how satisfying that would be! I hope somebody gets away with this someday.

RECOMMENDED READING

Your Ham Antenna Companion, by Paul Danzer, N1II, published by ARRL.
The ARRL Antenna Book (any recent edition), published by ARRL.
ARRL's Wire Antenna Classics, published by ARRL.
HF Antennas for All Locations, by L. A. Moxon, G6XN, published by RSGB, available from ARRL.
Practical Wire Antennas, by John D. Heys, G3BDQ, published by RSGB, available from ARRL.
HF Antenna Collection, published by RSGB, available from ARRL.
Smartuners for Stealth Antennas, SGC Inc, available for download at **www.sgcworld.com**. Featuring SGC auto-couplers, the 31-page book, in Adobe PDF format, is loaded with practical Stealth antenna ideas.

BIBLIOGRAPHY

Bowles, Chester, AA1EX, "40 Meters/40 Dollars," *QST*, Feb 1995, pp 51, 57.
Brogdon, Al, K3KMO, "A Modest Multiband Antenna," *QST*, Jul 1994, pp 68 - 69.

Cebik, L. B., W4RNL, "Invisible and Hidden Antennas," **http://www.cebik.com/a10/ant2.html.**

Ford, Steve, WB8IMY, "The Lure of the Ladder Line," *QST*, Dec 1993, pp 70 - 71.

Galgan, I. G., WA2VIA, "A Clothesline Dipole," *QST*, Sep 1991, p 32.

Kleinschmidt, Kirk, NTØZ, "Product Review: The MFJ-1786 High-Q Loop Antenna for 10 to 30 MHz," *QST*, Aug 1994, pp 62 - 64.

Lindquist, Rick, N1RL, "Product Review: Bilal Isotron 40," *QST*, Mar 1998, pp 73 - 74.

Holsopple, Curtis, K9CH, "Radical Radiators and Weird Wires," *QST*, Jan 1996, pp 61 - 63.

Monticelli, Dennis, AE6C, "A Simple, Effective Dual-Band Inverted-L Antenna," *QST*, Jul 1991, pp 38 - 41.

Morris, Richard, G2BZQ, "An In-Room, 80-Meter Transmitting Multiturn Loop Antenna," *QST*, Feb 1996, pp 43 - 45. (Feed-back, *QST*, May 1996, p 48.)

Muscolino, Bruce, W6TOY, "A Practical Stealth Antenna," *QST*, Jul 1995, pp 58 - 60.

Parker, Albert, N4AQ, "A Disguised Flagpole Antenna," *QST*, May 1993, p 65.

Schaffenberger, Paul, KB8N, "South Texas Stealth Sweep-stakes," *QST*, Nov 1995, pp 50 - 52.

Weaver, Mark, WB3BJF, "A Four-Band 'Tree' Vertical," *QST*, Nov 1995, pp 69 - 70.

Chapter 6

Close Quarters: Handling Interference

This chapter is about an annoying, unfortunate consequence of modern civilization—interference. EMI, RFI or just plain I. Unlike ham radio's formative years, we're swimming in a sea of RF and buried under mountains of questionably designed (regarding RF susceptibility, that is) and inexpensive electronic goodies (that we all enjoy). Way back when, hams interfered with TVs and telephones. Those problems still exist, of course, but today's scenarios can seem worse because of the sheer volume of stuff out there.

Doorbells used to be hard-wired. Now they run on batteries and RF. Garage doors used to be muscle-powered. Nowadays, they also rely on RF. VCRs weren't even on the scene a few years ago! Add wireless LANs, wireless speakers, baby monitors, cell phones, pagers, home automation systems, remote car starters—and who knows what else—and you'll only hit the tip of the iceberg.

As hams we also receive interference from some of these electronic marvels. Ironically, sometimes we even cause this interference to our own stations! I spent countless hours trying to keep my shack computer from spewing RF spurs up and down every band I could receive! After trying dozens of filters, conventional and unconventional RF enclosures and a host of other techniques I'm too embarrassed to repeat here, I solved the problem by simply using a different monitor! In my case, all of the 14- and 15-inch color monitors I hooked up caused big problems. Then I switched to a 10-year-old monochrome monitor—and found blissful radio silence! Why did this monitor work when all my other efforts had failed? Who knows! That's often the nature of these things. First, try the usual cures, which we'll discuss here. Then, if necessary, try something else (within reason)!

Actually, with a little luck and a lot of perseverance, most interference issues can be solved or minimized. Running low power and using good engineering practices will help your cause tremendously. I can only touch on a few common solutions in this chapter. If you're experiencing Stealth radio-related interference—from your rig or to your rig—I strongly suggest picking up a copy of *The ARRL RFI Book: Practical Cures for Radio Frequency Interference*. The RFI Book is larger than this entire book and is a comprehensive resource for fixing every imaginable interference problem in your home or mobile shack. It'll be the best $20 you've ever spent if RFI is threatening to ruin your enjoyment of ham radio.

DIPLOMACY

When interference rears its ugly head, who's to blame, anyway? Who is responsible for cleaning up the mess? The answers to these questions are many, and vary with the circumstances. If you're operating a properly licensed and engineered Amateur Radio station and you're fritzing out the neighbor's VCR, your buddy across the back fence may ultimately have to provide his own solution. It is, however, in keeping with common decency to at least point your neighbors toward possible solutions and to work with them as long as everyone's being friendly. If your wife, husband or mother is involved, however, you might bear the brunt of finding a solution, regardless of your excellent standing with the FCC!

Before we dig into specific problems and solutions, let's look at a few interesting facts from *The ARRL RFI Book*:

• Hams must operate their transmitters in accordance with all

appropriate FCC regulations. Make sure your station equipment is properly installed, has a good RF ground, uses a good low-pass filter at the station output, and so on. See **Figure 6.1**.
- Hams are not required to help their neighbors with RFI complaints that do not involve their transmissions (although they may elect to do so).
- The FCC considers telephones, VCRs, alarm systems, CD players, audio amplifiers (and so on) that receive RFI to be improperly functioning as radio receivers. The manufacturer should handle these design inadequacies.
- The RFI susceptibility of consumer electronic devices is limited only by the manufacturers' voluntary compliance with committee-developed standards, such as those of the Consumer Electronics Manufacturers Association. The voluntary standards do not address operating the equipment in close proximity to powerful transmitters. Transmitter operators are not responsible for RFI in such situations.
- In general, equipment owners are responsible for proper operation of their equipment. As an example, if your neighbors experience RFI from your properly licensed, engineered and operated Amateur Radio station, they are responsible for any corrective measures.
- FCC regulations require that ham transmitters not emit spurious signals that interfere with other radio services. This is the ham operator's sole regulatory requirement—and it doesn't apply to interference to non-radio consumer devices.

From a purely regulatory perspective, we're standing on pretty solid ground. If our transmitters are clean, interference is mostly their problem, not ours. In the real world we'll probably have to (or want to) be more accommodating. After all, operating a ham station in an apartment building with indoor antennas and a counterpoise wire for an RF ground may meet the letter of the law, but it may violate the spirit of the law. Besides, our Stealth stations might even interfere with our own consumer electronic devices, and in a typical Stealth setting we're likely to experience at least some RFI from those devices. Knowing how to handle RFI problems (and introduce others to possible solutions) is a good idea.

To keep things focused, from here on I'm assuming that any RFI problems are being experienced by you or other members of your household. I'll let you handle the neighbors on a case by case basis!

A LOW PASS TVI FILTER

Here's a low-pass filter that you can build. It's almost as good as commerical products. It should be adequate for all but the most severe TVI caused by transmitter harmonics.

Construction

The filter is constructed in an aluminum box measuring $3^1/_2 \times 2^1/_8 \times 1^5/_8$ inches. Input and output connectors are mounted at the center of each end. Use 5%-tolerance (2% would be better) capacitors. The 500-V capacitors specified should be more than adequate for a 200-W transmitter.

The coils are space wound from no. 18 enameled copper wire. Wind L1 and L3 using a $^1/_4$-inch drill bit as a form, use an $^1/_8$-inch bit for L2. To space the windings, wind two pieces of wire in parallel. To remove the extra winding, just grasp one wire and pull (with the windings still on the drill bit). Solder all parts to a two-terminal-with-ground solder strip as illustrated.

Tuning

Adjust the coils by spreading or squeezing the turns. Set L2 first, for maximum rejection of TV channel 2 (55.25 MHz). Then tune L1 and L3 for lowest SWR, or minimum insertion loss, at 28.5 MHz.

Here are two ways to adjust L2: Method 1 requires a grid-dip oscillator or solid-state dip meter. Temporarily short L2 (the end that connects to L1 and L3) to the grounded end of C2 with a *very short, fat* conductor. (A piece of tinned coax braid should do nicely.) Set the dip meter frequency by placing it near a TV set tuned to channel 2. Tune the dip meter to produce broad horizontal interference bars on the TV. Loosely couple the dip meter to L2, and adjust the coil for a dip at 55.25 MHz.

Method 2 requires a strong channel-2 signal and a TV fed with coax. CATV service should suffice. Connect the filter between the TV antenna connector and the feed line. Adjust L2 for maximum "snow" in the channel-2 picture. It may be necessary to temporarily short L1 and L3 to yield a strong enough signal.

Performance

The filter attenuates all VHF TV frequencies a minimum of about 48 dB. It has about 70 dB of rejection at the channel-2 carrier frequency (often one of the worst trouble spots). The worst-case insertion loss is about 0.3 dB, and the SWR is less than 1.3:1 below 29.0 MHz. The loss rises to over 0.4 dB at 29.7 MHz. If you operate at the high end of 10 meters, you may want to peak L1 and L3 at 29.0 MHz instead of 28.5 MHz.

Figure 6.1—Construction details for this homemade low-pass filter are taken from The ARRL RFI Book. *Commercial units can be found at your favorite Amateur Radio products dealer or at a hamfest near you.*

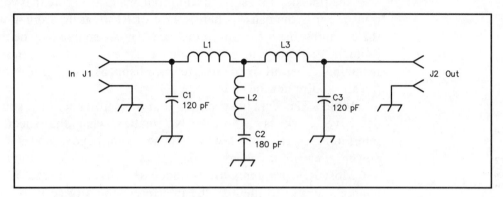

(A)

(B)

A low-pass filter for amateur transmitting use. The circuit (A) is constructed on a two-terminal-with-ground terminal strip (15TS003) which is fastened to the chassis with no. 4 solder lugs and hardware. (Part numbers in parentheses are catalog numbers from Mouser Electronics.) B is a photo of the completed filter.

$C1, C3$—120 pF, 5% tolerance, 500 V silver mica (232-1500-120).
$C2$—180 pF, 5% tolerance, 500 V silver mica (232-1500-180).
$L1, L3$—13 turns no. 18 wire, space wound $1/4$-inch ID.

$L2$—4 turns no. 18 wire, space wound $1/8$-inch ID.
$J1, J2$—SO-239 "UHF" connectors (16SO239) or RF connectors of the builder's choice.
Aluminum chassis—$3^1/_2 \times 2^1/_8 \times 1^5/_8$ inches (537-TF-773).

CAUSE *BEFORE* EFFECT

The first step in resolving RFI problems is to be sure that your transmitter is the cause. There's a lot of RF floating around out there—and some other transmitter or RF noise source may be the problem. Keeping a good logbook is excellent preventive medicine. After all, if the interference happens when you're off the air, you're not the cause.

If the RFI occurs in sync with your transmitter activity, some detective work is in order. No matter what the specific interference, perform a few tests to see what bands, modes and power levels are involved.

Most RFI problems *aren't* mode sensitive, although other people's ability to identify the interference and trace it to you probably is. Most people don't know Morse code, for example, and they can't recognize the deedle-eedle-eedle of RTTY. If they don't know that you're a ham and that you're on the air, chances are good that you won't be discovered as "the source." If you're blasting away on SSB, though, mentioning your name and QTH is probably a dead giveaway. Even if the "RFI audio" is distorted, they'll soon guess that it could be you.

Most RFI problems *are* power related. As I've mentioned before, running low power is a good way to eliminate or minimize potential RFI and RF exposure problems right from the start. For example, if your 50-W signal is tearing up the TV, try 10 or 25 W instead. If that clears things up you have the green light to safely operate until you can engineer a solution. Actually, you might not even need to get your hands dirty. Running 10 or 25 W instead of 50 W makes little or no practical difference on the receiving end.

Most RFI problems *are* frequency related. This can be a blessing. It helps in finding solutions, and it gives you an opportunity to work other bands while that solution is in progress. If 15 meters is the problem, work 40, 20 and 10 meters. It's not perfect, but it's a reasonable compromise.

WHERE TO START

The best place to start any Stealth radio RFI elimination plan is to:
- Use good engineering practices from the start.
- Run low power.

- Use the best possible antenna system (outdoor is better than indoor, higher is better than lower, and so on).
- Provide the best possible RF ground for your antenna system. (Use a counterpoise tuner or "artificial ground" if it helps.)
- Insert a high-quality low-pass filter at the output of your transmitter. The filter won't eliminate all types of RFI, but it will attenuate higher-frequency harmonics and spurious signals that might produce RFI. Using a low-pass filter is a low-cost, common-sense thing to do. If the FCC ever gets involved in an RFI issue at your address, using such a filter would certainly reflect favorably on you!

RFI BASICS

Remembering that this chapter is a mere introduction to RFI issues, there are several basic RFI categories:

- **Spurious emissions** (harmonics, mixing products, noise and other unwanted signals generated by your transmitter). Reducing transmitter power and using a low-pass filter can sometimes eliminate this type of interference. Let's say that your transmitter is damaged or misaligned and, while operating at 21.2 MHz, your station is producing a strong fifth harmonic near 106 MHz—right where your sister parks her FM broadcast receiver. Fixing the transmitter would likely fix the RFI problem, as would using a low-pass filter, which sharply attenuates any signals above 35 MHz or so (for a typical low-pass filter designed for HF operation up to 30 MHz).
- **Intermodulation and external rectification**. Poor-quality electrical connections (usually outdoors); corroded joints in downspouts, antenna towers or metal-sided buildings; bad solder joints, telephone systems and junction boxes and a whole host of similar items can radiate RF energy and harmonics when excited by your RF. These problems can be frustrating and difficult to track down.
- **Fundamental overload**. This is the most common culprit. Your transmitted RF (from your clean, perfectly engineered station) simply overpowers the affected device. What's more, your signal might be "getting into" the affected system in a variety of ways: antenna lead-ins, speaker wires, ac line cords, ground wires, or by other methods. In cases of "direct radiation pickup," an internal component might be receiving your RF signal directly! Fixing

direct radiation problems usually requires shielding the entire affected unit—which is rarely practical (it's time to talk to the manufacturer!). After all, it is difficult to watch a TV set that's sealed in a metal-lined box!

WHAT'S YOUR RFI MODE?

Without getting too far afield, your RFI cures will typically be aimed at *differential-mode* RFI or *common-mode* RFI. Differential-mode RFI involves a transmission line such as the coax that runs from a TV antenna to a TV receiver. If the TV antenna receives its desired TV signals and your undesired ham signal, it will pass both signals on to the TV receiver through the coax (and the ham RF may interfere with the set). If you install a high-pass filter at the TV receiver's antenna terminals—a common differential-mode RFI cure—the filter will attenuate the lower-frequency ham RF while passing the desired TV RF.

Differential-mode cures can be simple. Unfortunately, most RFI is a common-mode problem, where the interfering signal is arriving via both conductors (the coax center conductor and the shield braid) or all conductors (antenna leads, ground leads, speaker wires, ac line cords, dc power cables, etc).

As maddening as it may be, determining your specific interference modes may be necessary, as differential-mode cures won't work for common-mode problems, and vice-versa.

A FEW SPECIFIC CURES

As you read through the scenarios that follow you'll notice a common pattern of signal sleuthing. When RF from your station is adversely affecting some other device, the first step in solving the problem is determining exactly how the RF is "being received." Just because a TV set has an antenna, don't assume that the unwanted signals are getting in through the front door. Other wires in the system can easily receive RF. These include power cords, speaker wires, audio/video input and output cables, ground leads, and so on. Disconnecting the various "potential RFI antennas" is a good way to start the tracking process.

TVs and VCRs

When it's time for whatever favorite program around your neighborhood, and you're messing up the picture, you're going to catch an earful from the nearby viewing public! (Nowadays this often results from signals getting into the TV tuner built into a companion VCR.) To solve this problem, let's start with the easiest steps first.

Disconnect the coax or twin lead from the set's antenna terminals (or the antenna input on the VCR) and try a few test transmissions from your shack. If the RFI stops you know that the problem is in the antenna side of the system and not in the power leads, speaker wires or interconnecting cables. If the problem is being caused by harmonics of your transmitted signal or simple front-end-overload (differential-mode RFI), installing a high-pass filter — shown in **Figure 6.2** — at the set's antenna input (and/or a low-pass filter at your transmitter output) may be all that's necessary.

These filters are inexpensive, and are available at most TV repair shops and electronics parts stores. A small in-line filter is even sold at RadioShack stores. Be sure to buy the correct version (75-Ω coaxial cable or 300-Ω twin lead). Because modern ham equipment is generally excellent when it comes to harmonics and other spurious signals, further steps may be necessary. In fact, if a high-pass filter doesn't completely cure your RFI problem, leave it installed as you try other remedies. It certainly won't hurt!

If the interfering signal is still a problem when a high-pass filter is in-line—or if the interference is present when the antenna is disconnected—unwanted RF is entering the system through the outside of the antenna lead-in, the power cable or some other interconnecting cable. This is usually common-mode RFI. If the set has A/V cables or speaker wires running to a stereo amplifier or other home theater components, disconnect these lines to see if the RFI situation changes. If it does, plugging them back in one at a time will often pinpoint the source of the problem. (Unless the set has internal battery power, disconnecting the power cable isn't practical!)

Figure 6.2—High-pass TVI filters handle differential-mode interference in antenna-connected TV and FM broadcast receivers.

RadioShack, your local ham radio store and various mail-order catalogs sell ferrite cores in several shapes and sizes to help you in your plight. My tongue-in-cheek title for this chapter is "Ferrite is Your Friend." Cleaning up my own shack's RFI problems required quite a few cores! Treating signal cables, coaxial antenna leads, speaker wires and power cords for common-mode RFI requires similar measures, so don't be shy about applying them to the ac line cord, too.

To make a common-mode RFI choke, wrap a few turns of the cable, cord or wire through an appropriately sized ferrite core as close to the chassis/connector end as practical, securing the windings with electrical tape if necessary. This will often reduce or eliminate the RFI and will let you know whether you're on the right track. Curing severe common-mode RFI may require chokes on several cables or interconnects (ac power, antenna and A/V inputs, for example).

The effectiveness of common-mode chokes is often proportional to the number of turns through the ferrite core. Winding several turns of speaker wire through a toroidal core is easy; winding a similar number of turns with bulkier coaxial cable isn't! To make things easier, look for square "split cores" that open and close with plastic retainers. Open the core (splitting the halves), wind as many turns as practical, close the core and secure the windings with tape. See **Figure 6.3**.

If your RF signal is being received and conducted via the ac power wiring, a commercial ac line filter may be what you need. If winding the ac power cord onto a ferrite core helps (but doesn't completely cure) the situation, an outlet-mounted line filter may finish off the problem. These filters are also avail-

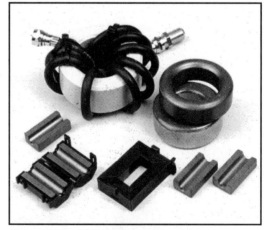

Figure 6.3—This photo shows several styles of common-mode chokes.

able from RadioShack, Amateur Radio equipment dealers and catalog vendors.

There are many other potential causes for TV and VCR RFI problems. If you live in a fringe area and use a large tower-mounted TV antenna, chances are good that you have a tower- or mast-mounted antenna preamplifier. These units are known for their susceptibility to strong signals—in band and out! The same goes for many inexpensive 75-Ω cable TV line or distribution amplifiers.

If your in-house cable or cable distribution system uses signal splitters, make sure all unused ports are terminated with screw-on 75-Ω terminators (screw-on resistors) as shown in **Figure 6.4**. Leaving unused ports open can disturb the balance or the impedance of the normally closed coaxial system and let the RFI demons in through the back door.

Stereos and Computer Sound Systems

As you can imagine, solving these RFI problems involves techniques identical to those listed above. To determine whether the tuner portion of the system is experiencing problems from harmonics or front-end overload, disconnect the antenna. If the interference disappears, install a high-pass filter at the antenna terminal (50-Ω for coax, 300-Ω for twin lead).

If the RFI is still present, leave the high-pass filter in place and begin the search for common-mode culprits as

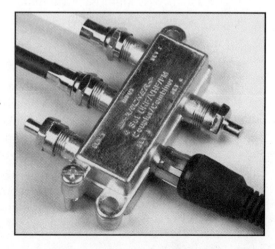

Figure 6.4—The unused ports of all multi-port splitters should be properly terminated with screw-on 75-Ω terminating resistors.

described earlier. Disconnect cables and speaker wires and reconnect one at a time to pinpoint the trouble spot(s) and apply/wind common-mode chokes as necessary. Speaker wires are often cut to convenient lengths that happen to correspond with quarter-wavelength ham antenna dimensions. This may make them more susceptible to receiving RFI at certain frequencies. The speaker wires are acting as tuned antennas! Lengthening or shortening the speaker leads can sometimes help eliminate or reduce RFI.

For speaker-related problems, RF signals often enter the system via the speaker leads, which conduct RF energy to diodes or transistors in the audio amplifier circuits. The solid-state devices rectify the RF and mix the distorted signal into the amplified audio chain. Adding common-mode chokes often keeps the RF from reaching the amplifier circuits. RFI-proofing older tube-type audio equipment sometimes involved connecting capacitors across the speaker terminals. Don't try this with modern solid-state gear! In fact, don't try any other questionable solutions without asking an experienced helper or reading an appropriate article or book on the subject!

Telephones

What would we do without telephones? If your radio is on its way to completing its Worked All Phones award, you already know. When the telephone is rendered useless because of RF, other people will, at a minimum, screech, yell, threaten or otherwise complain loudly!

Because telephones, answering machines, voice-mail systems, caller ID units, autodialers, computer modems, speaker phones, cordless phones and fax machines are everywhere nowadays, telephone RFI is a growing concern. To make matters worse, today's devices are loaded with solid-state circuits that are just begging to rectify your RF (see the previous section on audio amplifiers and speakers).

The most common way to clear up RFI that is being "received" via telephone lines is to install in-line filters or common-mode chokes at the service entry, at each telephone, and sometimes in the handset lines! These filters, shown in **Figure 6.5**, are available from mail-order vendors, retailers and some-times from the phone company.

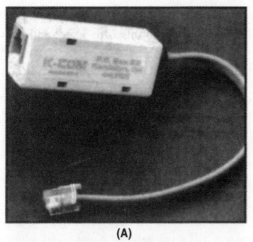 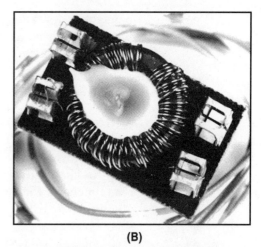

(A)   (B)

Figure 6.5—Telephone RFI filters. A is a commercial unit with modular connectors. B is a compact, common-mode choke that is inserted in the telephone line. A dab of silicone adhesive through the center of the toroid core holds the choke to a piece of insulating board with clips for easy connection to the phone line. You can make a similar choke by winding a bifilar pair of wires onto a $1/2$-inch OD, mix 75 ferrite toroidal core. Twist two no. 30 enameled wires together until you have about five to ten twists per inch, and then wind 25 turns to cover about 80% of the circumference of the core. Use different colored enamel wires, or use an ohmmeter to check for continuity through the wires after the toroid is wound, then mark the ends so you can insert it in the phone lines properly.

If components inside the telephone are receiving your RF directly, try reducing power and/or moving your antenna farther from the telephone, or just get a different phone. If you can do without a modern convenience or two, pick up a 1970s-era touch-tone phone at the next flea market. Older phones often perform better in RF-rich environments.

INTERFERENCE TO YOUR STATION

There's a lot of RF garbage floating around out there, and sooner or later it will make its presence known in your shack. Common RF-noise generators include motors, drills, saws, garage door openers, furnace blowers, buzzers, thermostats, car ignitions, touch lamps, dimmers and more.

If these devices are in your house you can track them down and deal with them (maybe). RF garbage can propagate quite handily, so you may have to live with it. After all, if you use your handy direction-finding rig to track a noise problem to the neighbor down the street, you'll probably have to divulge the fact that you're a ham—which is usually counterproductive if Stealth operation is a concern.

Most high-end radios have two types of noise blankers for dealing with this stuff. One is for pulse-type auto ignition noise while the other is for "wider" noise sources. Because designing effective noise blankers is a black art, chances are good that the blanker(s) in your rig will leave something to be desired.

If noise is a real problem at your QTH, consider inserting an up-to-date digital signal processor (DSP) unit in your station's audio chain. I use a Timewave 599ZX, shown in **Figure 6.6.** In addition to having razor-sharp audio filters, the 599ZX often helps tame "random" noise makers in my vicinity—even computer hash. DSP boxes don't always help, of course, but when they do help, they usually help a lot.

COMPUTER NOISE

Many shacks have at least one personal computer. Therefore, many shacks suffer from some form of "PC RFI." With multiple square-wave oscillators (and multipliers) shrieking from dc to daylight, personal computers can really spew RF—especially the nasty buzz-saw noise that wipes out entire regions on your radio dial.

Taming these ever-so-handy beasts involves keeping the RF junk inside the computer case and away from your radios and antennas. To that end you can use PCs that have tightly sealed metal cases; keep slot covers in place over every open rear-panel slot and wind

Figure 6.6—Timewave's Model DSP-599ZX is representative of today's "do everything" external DSP boxes. Although they're useful for a wide variety of day-to-day operating tasks, DSP filters can sometimes tame noise sources that can't be affected any other way. (author photo)

common-mode RF chokes onto interconnecting cables as necessary (keyboard, mouse, modem, keyer, video, ac power, and so on).

Keeping a healthy distance between your PC and your antenna is often an easier and more effective solution. The problem is, especially for Stealth installations, it's not always possible. My Stealth antennas have always been very close to my operating positions, and computer noise has plagued me for years. My only solution was to use an old, slow laptop computer. It didn't smear the bands with RF crud, but it didn't run all of my ham radio software, either.

As I mentioned before, I finally got things under control. As long as I didn't have a monitor connected to my shack computer, the setup was RF quiet everywhere. Once the monitor was hooked up, however, the bands were almost wiped out. I tried severaL color monitors to no avail. When I connected an older monochrome unit, however, the bands remained in fine shape. I can still hear a little junk here and there, but the improvement is like night and day. The moral of this story? Parts substitution!

Because there are hundreds of PC parts manufacturers the world over, some monitors, some keyboards, some mice, etc, are much quieter than others when it comes to radiating RF. If you're nearing your wit's end, swap a few parts, especially external, cable-connected parts. You, too, may be pleasantly surprised. I only wish I had made my monitor swap years ago!

As this chapter comes to a close, let me again plug *The ARRL RFI Book*. Every time I read the thing I learn something I wish I had known years ago. Do yourself a favor, Stealth Operator. Get your own copy and save yourself a lot of grief.

RECOMMENDED READING

Hare, Ed, W1RFI, editor, *The ARRL RFI Book: Practical Cures for Radio Frequency Interference*, ARRL, 1998.

Buffington, Roger, AB6WR, "Real Fun Interference (RFI)," *QST*, Feb 1994, p 78.

Ford, Steve, WB8IMY, "TVI, CATVI and VCRI," *QST*, Mar 1994, p 72.

Krieger, Pete, WA8KZH, "Basic Steps Toward Eliminating Telephone RFI," *QST*, May 1991, p 22.

Trescott, Max, K3QM, "Basic Steps Toward Tracing and Eliminating Power-Line Interference," *QST*, Nov 1991, p 43.

Chapter 7

Radio in Motion

If you think that the heydays of mobile hamming peaked in the '60s, '70s and '80s, you're way behind the times! The radios of that era are gigantic and kludgy when compared with today's tiny, do-everything mobile rigs. As recently as the mid-'80s I balked at the thought of somehow strapping my Kenwood TS-430S HF transceiver into my Japanese hatchback. Although quite compact for its day, the '430 is a whopper compared to the amazingly small mobile rigs that are now available from Alinco, ICOM, Yaesu, Ten-Tec, SGC and others. (Old-Timer mobile hams did have one advantage: Big metal car bodies that made antenna installation a breeze. Of course, they needed all that extra space to house those big ol' radios!)

Thanks to surface-mount technology and digital electronics, mobile ham radio is peaking *right now*—and tomorrow will be even better. I don't know what you've experienced, but I've worked more mobile stations on HF and 6 meters in the past year or so than I have in the previous 20 years combined. Mobile operators are coming out of the woodwork!

And it's a good thing, too. The present mobile craze is a real benefit to Stealth operators who endure antenna restrictions, RFI

problems or an inability to set up an indoor shack of any kind. If you find yourself in one of those situations, don't worry. All is not lost—if you have a car, that is!

Mobile hamming also benefits traveling hams and people who are busy, busy, busy. Ham radio can be a constant companion whether you're making a transcontinental run or a daily commute to the office. Contesting, ragchewing, DXing—whatever you enjoy. Today's mobile hams are in no way second-class ham citizens.

MOBILE BASICS

Before we get into the specifics of various installations, let's look at mobile hamming at a basic level.

In earlier chapters we learned about balanced and unbalanced antennas and various types of RF grounds. Remembering the basics, the vast majority of mobile ham stations use vertical antennas worked against the body and frame of a car or truck, which serves as the RF ground. As with ground-mounted verticals, the size of the radiating element and the quality and size of the RF ground determine the overall effectiveness of the antenna system. Big radiators and big RF grounds mean big signals, and vice-versa.

Unless you drive a vehicle the size of an aircraft carrier, this imposes certain realities and restrictions. At 2 meters and up, where car bodies are more than a wavelength in size and whip antennas are a quarter-wavelength or longer, mobile installations are typically efficient and represent little or no compromise when compared with typical base station antennas mounted at the same height.

At HF, however, efficiency goes downhill in a hurry. At 80 meters, for example, car bodies are a fraction of a wavelength, as are mobile whip antennas. No matter where you mount them, physically small antennas with poor RF grounds don't radiate powerhouse signals.

This doesn't mean that 80-meter operation is impossible, but it does mean that you have to pay attention to all of the details if you want to enjoy operating "down there." A well-installed 80-meter mobile antenna (8 to 12 feet tall!) with good ground connections is about 9% efficient. A 5-foot-long whip with poor ground connections is about 1% efficient—it's essentially a "leaky" dummy load! When connected to your 100-W transceiver (all other factors ignored), the two antennas will radiate about 9 W and 1 W, respectively. You'll still make contacts with both setups, but there's no escaping the fact that the 9% station will handily outperform its 1% cousin.

The good news is that antenna performance increases with frequency. If 80 and 160 meters represent worst-case scenarios—and they do—mobile hamming from 20 through 10 meters is a relative breeze. On the higher bands antennas are much closer to being at least a quarter-wavelength in size, and the RF ground performance of typical car bodies improves markedly. So, if you want an easy introduction to mobile hamming, start out on the upper HF bands or any VHF/UHF band and leave the low-frequency stuff for later, after you've logged some miles as a mobile ham.

BANDS AND MODES

Working the world from your car is a lot like working the world from home. The bands open and close, noise levels vary from quiet to thunderous, pileups still take work to crack, and so on. You won't have as much equipment (I hope!) and your operating flexibility will be tempered by the fact that you're driving a moving vehicle. You'll also be using a compromise antenna, so your "command" of the bands may be limited. Here's a brief rundown on each band from a mobile perspective.

160 meters: Mobile work here is best left to antenna experimenters and operators who pack big amplifiers and even bigger antennas. You can make QSOs here, but 160-meter mobile operation is far from casual.

80 meters: This is a great nighttime band, but 80 can be crowded and difficult for mobiles. If you want a big signal you'll have to use a *BIG* antenna. Screwdriver-type antennas (such as TJ Antennas' BB3 Broadbander) or giant center-loaded whips with massive coils (such as the imposing Texas Bugcatcher)—both of which top out at 8 to 12 feet—are required. Small, sleek, compact or Stealthy whips will disappoint you here. Some 80-meter mobile specialists use 500-W dc-powered amplifiers.

40 meters: Forty isn't as demanding as 80 but it's still more difficult to put out a good signal here than it is on the higher frequency bands. You'll have steady regional coverage during the daytime, with longer-distance QSOs at night—if you're not wiped out by shortwave broadcasters, that is.

30 meters: Because it's a CW/digital only band, you'll find few mobiles here. If you're one of the select few, however, 30 meters is constantly open to somewhere, uncrowded and almost always pleasant.

20 meters: Twenty is crowded with high-power base stations, but it's also crowded with people to chat with and is a great band for mobiles. Smaller, less obtrusive antennas work well here. Unlike the higher frequency bands, 20 meters is usually open to somewhere, even into the night.

17 meters: Arguably, 17 meters is the best band for mobile HF. Afternoon openings are excellent, the band is uncrowded and a little signal goes a long way.

15 meters: Propagation here is similar to 17 meters but the band is open less often. At sunspot peaks, 15 meters is a solid DX producer in the afternoon and early evening.

12 meters: Okay during peak sunspot years; supports less activity than some of the other HF bands during off years.

10 meters: Near sunspot peaks 10 meters is wall to wall with signals. Ten watts and a mobile whip can make you king of the band. Even during the doldrums, 10 meters opens periodically thanks to sporadic E and other less-common propagation modes.

6 meters: Mobile activity on 6 meters is booming thanks to a flood of compact HF/VHF mini mobile rigs. You won't work stations here everyday (unless they're locals), but during sporadic-E season, you can regularly hear and work mobile stations on 6 meters.

As you can imagine, the most common mobile modes are SSB (HF) and FM (VHF and up). But don't count Morse code out just yet. A small but active cadre of mobile CW operators can often be heard, especially on 40 and 30 meters. If you're a solid CW operator, give mobile CW a try. You'll have a period of adjustment, to be sure, and you'll have to engineer a place to mount your keyer paddle. If you think talking on cell phones causes accidents, just imagine how much more engrossing/distracting mobile CW might be! Be careful and be safe. (Maybe you can get someone else to do the driving.)

EQUIPMENT

Because mobile hamming isn't exactly new, just about any radio that can be powered by 13.8 V dc can be made to operate in a mobile environment. Unless you're driving a big truck or an older, bigger car, chances are good that you'll have better luck with a newer, smaller rig designed for mobile service.

Mini mobile rigs from Alinco, Yaesu, ICOM, Ten-Tec, SGC and others have swelled in popularity. Many of these all-mode rigs cover HF, VHF and even UHF bands with digital readouts and lots of bells and whistles. About the size of a cigar box, these rigs are easy to find space for in a crowded dashboard. All have conventional mobile mounts and some have detachable "control heads" that allow the bulk of the radio to be mounted under the seat or in the trunk. See **Figure 7.1**.

INSTALLATION

Before plunging in, be sure to plan your installation thoroughly and read up on the subject if you're inexperienced. Heck—read up on mobile operating even if you *are* experienced!

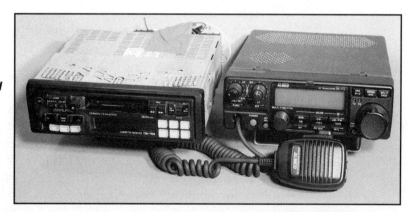

Figure 7.1—My Alinco DX-70 is at home in the car or in my shack. One reason is its size. When they are placed side by side, the car stereo at the left seems rather large! (author photo)

My favorite book on mobile operating is *Everything You Forgot to Ask About HF Mobileering*, by Don Johnson, W6AAQ. As the title implies, Don has forgotten more about the subject than most of us will ever "know." At press time the book costs $8 and is available from Worldradio Books in Sacramento, California (or your favorite ham radio dealer). If you're serious about mobile hamming, do yourself a favor and get a copy *before* you begin!

Where to mount the radio depends on its size, whether it has a detachable head, the type of vehicle it's being installed in, and so on. Older rigs require more space and more robust mounts. Newer rigs are tiny and lightweight. I use a home-brewed velcro mount and quick-disconnect cables to allow me to quickly remove the radio for storage or service in my home shack.

The exact routing of the power and antenna leads also depends on a lot of variables—but it has to be done, of course. Finding a way through the firewall probably won't require any drilling. In most cars and trucks you can pop out a grommet or piggyback on the same hole used by a hose, cable or wire. If you do have to drill, mark your hole carefully and don't drill any deeper than necessary to avoid hitting wires, hoses, ducts or anything else.

Whatever your rig, whatever your vehicle, your first tasks are to:
- Mount the radio securely and position it to allow easy operation while underway. Stay away from the airbags!
- Run a pair of *heavy-gauge* power supply wires from the cockpit-mounted rig *to the vehicle battery*.
- Find a sturdy, low-resistance chassis ground point *as close to the rig as possible*. This transceiver ground is critical—don't ignore it.
- Run a coaxial cable from the rig to the antenna mount.

POWER CABLES

The power supply wiring, even for 100-W transceivers, should be *beefy—ridiculously beefy* if possible! Because the run is short, splurge on well-insulated, flexible power leads that are 8-gauge *or larger*. "Extra-rubbery" arc welding cables are ideal (in size 0-0). Using monster-size cables will practically eliminate voltage drop and minimize RFI and noise pickup under the hood. It's not intuitive, but experience has shown that huge power

Figure 7.2—Be sure to properly fuse both power leads at the battery.

supply cables can pay huge dividends in reduced noise and RFI. Because the mobile environment is "radio hostile," take advantage of every opportunity to improve the odds. Use a few inches of smaller-gauge wire to make the connections at the back of your rig.

Keep the power supply wires away from the car's wiring harness, hot engine parts and rotating fan blades. And make sure the access hole through the firewall is properly insulated, preferably with a rubber grommet. With vibration, temperature changes and shock, the edges of a sharp, bare hole can saw through the power lead insulation and short your vehicle battery to ground.

As shown in **Figure 7.2**, the positive and negative supply cables should *each* be fused *at the vehicle battery*. Without both fuses you needlessly risk a nasty electrical fire. With the full power of a charged car battery behind them, a shorted power cable could easily start a fire. Use two correctly sized fuses and be safe.

Use clean, new, properly sized battery terminals when making your power supply connections. Don't jury rig connections that might pass more than a hundred amps of dc!

GROUNDS, PART ONE

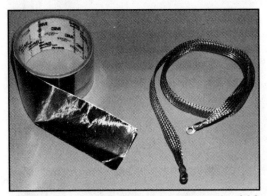

Figure 7.3—Low-inductance copper braid is used to make effective vehicle ground connections.

Good grounds are critically important in mobile setups. The first ground you need to establish is right at the rig itself. Don't rely on the negative power supply lead or the shield braid of the coaxial antenna feed. Connect a low-inductance braid or strap (shown in **Figure 7.3**) from the transceiver chassis directly to the nearest piece of frame metal. Drill a hole and install a ground bolt in the firewall, in the metal dashboard frame, the floorboards or the vehicle seat mounts. Don't use a long, meandering ground wire—just make it big and make it direct.

COAX

Let's talk about coax for a minute. We obviously want to use the good stuff—but what *is* the good stuff when it comes to mobile installations? First, buy new coax and forget about recycling that length coiled up in the tool shed. Buy cable that has a sturdy, non-contaminating jacket; a 95% (or greater) shield braid; a solid dielectric (avoid foam); and a stranded center conductor. RG-8MX, RG-213 and flexible LMR-400 should do nicely.

What's the proper length for your coaxial cable? Whatever length is required to span the run from your rig to your antenna. Don't worry about cutting the coax to any magical length.

ANTENNAS AND MOUNTS

Mobile antennas come in a variety of shapes, sizes and price ranges. Which antenna you use depends on your interests, your favorite bands and your vehicle. Antenna choices also depend on just how much metal you (or your spouse) want to bolt onto your car or truck!

Exactly where to mount your mobile antenna is the first order of business. As with any antenna, from a performance perspective, higher is better. (Unfortunately, from an aesthetic point of view, lower may be *better looking*.) Performance-wise, mounting the antenna in the center of the vehicle roof is best. The trunk and hood decks are next. And the front or rear bumpers are the poorest places to mount antennas. (Giant center-loaded whips with long bottom masts probably have to be mounted to a bumper or low on the body for mechanical reasons.) See **Figure 7.4**.

Antenna mounts come in all shapes and sizes. Some bolt directly to your car body, some attach to the lip of your car's trunk and some use powerful magnets to hold your antenna in place—but they all simply provide a place to mount your whip antenna. As with any backyard vertical, the whip must be insulated from the car body and the coax shield must be connected directly to the car body/frame. (See Grounds, Part Two, later in this chapter, for more information.)

If you have to use a bumper or trailer hitch mount, raise the base of the antenna by adding a mast tall enough to elevate the antenna's center loading coil at least a foot above the roof of the vehicle. If you don't, ground losses will be excessive.

There are a lot of mobile HF antennas to choose from, but because of space restrictions, we're only going to discuss three types

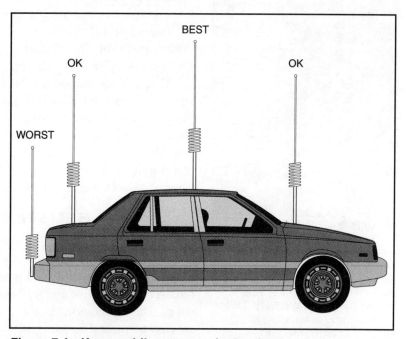

Figure 7.4—Keep mobile antennas in the clear!

here. Before we get specific, however, here are a few mobile antenna tips. Doing a little extra research on this topic can really pay off.

- Don't use antenna mounts with springs. The springs may look nifty and they may help preserve your antenna when you drive into a parking garage, but as the coils flex and rub together they create clicks and other unwanted electrical noise.
- Big antennas generally work better than their smaller cousins.
- Resonant antennas almost always work better than non-resonant antennas that are fed via tuners or tuning networks. Resonant antennas "take power" naturally while non-resonant antennas must be "force fed." So, if your 80-meter antenna is trimmed for 3.95 MHz and you want to operate at 3.8 MHz, retune the whip or the antenna's loading coil to bring it to resonance at the new operating frequency—don't try to fudge with the feed point transformer or "tweak" the mobile antenna with an antenna tuner.
- Avoid antennas that use small-diameter loading coils, loading coils made from small-diameter wire and loading coils that have close-wound turns. Coils, like mobile power leads, should be beefy.
- At the risk of generalizing, if your mobile whip has a loading coil it should be located somewhere near the center of the antenna. Forget base-loaded whips—they're too inefficient.

I've already harped on this before, but it's worth repeating: Never use a cockpit-mounted antenna tuner to "match" your mobile whip. Mobile antennas are physically small and offer compromised performance at best. Don't waste precious decibels by using a tuner! If you require *extreme* frequency agility, place the antenna tuner or autocoupler at the antenna feed point.

HAMSTICKS

If you want a simple, inexpensive and reasonably effective "starter" antenna, or if you want to operate on a single band at 20 meters or higher frequency, consider Lakeview's attractive and well-made line of monoband whips. Dubbed *Hamsticks*, these helically loaded whips are easy to use and easy to tune (see **Figure 7.5**). Models are available for every band from 80 through 6 meters.

Figure 7.5—This 20-meter Hamstick adorns a car that's driven by Steve, WB8IMY. This mag-mount installation is simple, uncluttered and effective.

At press time, *Hamsticks* are selling for $25 each. They are compatible with a wide variety of standard mobile antenna mounts. Performance on the low bands isn't stellar (a simple matter of physics), but from 20 through 6 meters, *Hamsticks* hold their own and have an impressive cost/benefit ratio!

SCREWDRIVER ANTENNAS

As mentioned previously, resonance is a key issue when discussing mobile antennas, especially antennas for 80 and 40 meters. A typical mobile whip is resonant at one frequency. At or near that frequency, SWR stays in an acceptable range and because the antenna is resonant (or nearly so), the antenna performs to the best of its ability (still a considerable compromise when compared with full-size antennas).

When we need to move to a frequency that's outside the antenna's "acceptable SWR" window (or need to use the antenna on another band) we run into a problem. If we use a cockpit-mounted antenna tuner the antenna's efficiency—already way too low—falls dramatically. We can stop the car and re-tune the antenna for a different part of the band, or, with some designs, change a jumper or coil tap. But stopping and tuning is inconvenient at best, and most antennas that use jumpers and tapped coils suffer performance problems because of coil Q and other esoteric factors. What we need, then, is a resonant mobile antenna that can be adjusted from the operating position inside the car. Enter the Screwdriver antenna!

Ostensibly invented in the early '90s by Mobile HF guru Don Johnson, W6AAQ, screwdriver antennas use reversible electric motors commonly found in cordless screwdrivers to rotate an adjustable center-mounted coil. Instead of using a jumper to short turns of an open-air coil (lossy), screwdriver antennas extend or retract the coil like the threads of a screw. This shorts all or part of the center loading coil, resonating the antenna system (which includes a sturdy bottom mast below the coil and a whip, or "stinger," that's attached to the top of the coil) *anywhere* from 3.5 to 30 MHz.

To tune the antenna to resonance, operators use receiver band noise, dash-mounted SWR or field-strength meters (models that emit tones or beeps work best) or inexpensive automatic tuning accessories (available as kits or fully assembled).

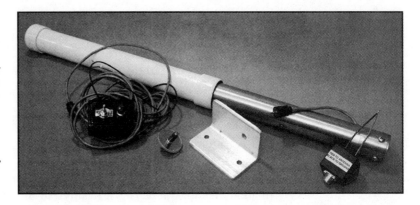

Figure 7.6—TJ Antenna Co calls its screwdriver antenna the BB3 Broadbander. The remotely tuned resonant whip covers all frequencies between 3.5 and 30 MHz. This one is ready for installation. Not shown is the whip or "stinger." (author photo)

Screwdriver-type antennas are available from several manufacturers. The one shown in **Figure 7.6** is my BB3 Broadbander, made by TJ Antenna Co of Farmington, New Mexico. The antenna is currently "between vehicles." When my truck comes back from the body shop with a fresh coat of paint, the BB3 is going on the back bumper (with a bottom mast to elevate the loading coil).

Over the past few years, screwdriver antennas have consistently scored at the top of the pack at dozens of mobile antenna "shootouts" from coast to coast. They cover all bands from 80 through 10 meters (and the shortwave bands in between), and are tuned from the operating position.

If you can get away with mounting a big, beautiful HF antenna on your car or truck and need multiband coverage (or a high-performance antenna for 80 or 40 meters), a screwdriver is just what the doctor ordered.

AUTOCOUPLER-FED WHIPS

Resonant screwdriver-type antennas offer superior performance on all bands, but they tune somewhat slowly. Fidgety "band hoppers" (the ham equivalent of channel surfers) find them tedious. If you need to operate *here, there and everywhere* as you drive down the road, nothing matches the flexibility of a big whip fed from a feed-point-mounted autocoupler.

Alinco, ICOM and other companies that manufacture today's mini mobile rigs make feed-point-mounted autocouplers (functional but somewhat slow and limited in matching range), but SGC's Model 231 autocoupler, shown in **Figure 7.7**, is the present quick-draw champion. It will match a large mobile whip on just about any frequency from 160 through 6 meters in the blink of an eye. As an additional benefit to busy drivers, tuning requires no buttons, switches, meters, or knobs. Simply key your rig (which sends RF to the coupler) and the tuner does the rest—fast. Band changes are virtually instantaneous because the tuner memorizes the tuning solutions required to match your antenna on all of your favorite bands.

Before you get carried away, there are a couple of catches. Although your whip may be resonant at one or two frequencies (unlike the screwdriver, which can resonate anywhere), it's *non-resonant* everywhere else. The autocoupler can match the load, but autocoupler-fed whips aren't as efficient as resonant antennas. They're fast, flexible and frequency agile, but they suffer from reduced efficiency, especially on the low-frequency bands. On 20 through 6 meters, however, where a big whip is more like a quarter-wave vertical (or more), feel free to QSY at your convenience.

Mounting can also be an issue. The autocoupler, about the size of a big dictionary, must be mounted at *or very near* the antenna feed point. In other words, the antenna *starts* at the coupler's output terminal and you can't run *any* coax from the

Figure 7.7—This is the business end of SGC's SG-231 autocoupler (no pun intended). Your vehicle's chassis ground connects to the unit's back plate while your mobile whip connects to the insulated post. Similar models are available from several manufacturers. (author photo)

tuner to the antenna. If you mount the coupler in the trunk, for example, you'll have to run at least a few inches of the antenna *inside the trunk*. If you mount the coupler on or under the vehicle, weather and theft issues arise (the 231 is designed for outdoor installations). SGC makes a quick-mount mobile tuner/antenna combo, but because it's primarily sold to military and government buyers, it's too expensive for the vast majority of ham users.

For mobile operators with portable tendencies, autocoupler setups can be especially handy. When you arrive at your field destination, simply toss a longer wire into the trees and unroll a radial or two (I recommend an inverted-L configuration) and you have an instant, high-performance portable antenna that can be quickly tuned on all bands.

GROUNDS, PART TWO

Good RF grounds are critical for mobile HF performance. We've already talked about grounding the rig in the cockpit. Now it's time to talk about grounding things at the antenna feed point and elsewhere.

Despite comments from well-meaning bystanders, every mobile antenna mount must have a good, low-impedance connection to ground—and that includes magnetic mounts! Without that connection to the vehicle body/frame, your RF performance will suffer tremendously.

QST Managing Editor Steve Ford, WB8IMY, found that out when he started operating HF mobile. **Figure 7.8** shows Steve's problem and its solution. Figure 7.8A shows WB8IMY's unmodified HF mag-mount. Although the mount easily retained

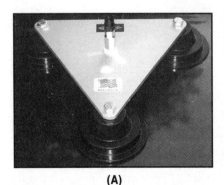
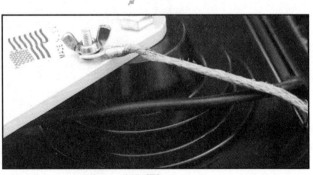

(A) (B)

Figure 7.8—The factory-fresh mag mount shown at A was improved by adding a wing nut and a low-inductance ground connection, as shown at B.

Figure 7.9—A short piece of metal can turn a trailer hitch into an effective and sturdy antenna mount.

his 20-meter Hamstick, it didn't provide a ground connection to the vehicle body/frame. Figure 7.8B shows Steve's solution—a piece of copper braid that connects the antenna mount's base plate to the inside trunk lid of the car.

Bumper mounts and other direct-to-the-frame mounts, such as the handy setup shown in **Figure 7.9**, usually provide an *okay* connection to ground—but even direct frame/body connections can usually be improved with the addition of a ground strap made from copper braid or strap.

The bumper mount, for example, relies on the metal-to-metal contact between the steel mounting prong and the steel trailer hitch. In turn, the trailer hitch is bolted to the steel frame of the car. Starting at the antenna mount, that's steel to steel to steel to steel. Steel isn't the lowest-resistance metal on the chart—and who knows how bad the rust and the crud might be between some of the pieces, especially those under the car...

Jumpers made from copper braid or strap, held in place by large sheet metal screws (or bolts, if possible) and star washers, could improve those connections and reduce the resistance between the various parts and connections along the way. It might sound excessive, but reducing the resistance in the ground connection by even a small amount can dramatically improve antenna performance and help reduce or eliminate noise (more on that in the next section). The bottom line is, you can never have a ground connection that's *too good*.

NOISE AND RFI

Compared to most home shacks, mobile installations are pretty hostile. Your rig may encounter temperature extremes, voltage swings, mechanical shock, excessive humidity and a host of other factors that aren't relevant at home. You may also have to deal with electrical noise from the ignition system, on-board computers, blowers, fans, defrosters and so on. You may even cause interference *to* your car's electrical system!

W6AAQ's book covers noise and RFI in detail, as does *The ARRL RFI Book*. If the basic steps detailed here don't take care of your problem, be sure to follow up with more in-depth advice.

The first step in troubleshooting pesky noise problems is to determine where the noise is really coming from. If your radio is installed and functional, disconnecting the antenna cable will help you determine if the noise is being "received" by the antenna system or entering the radio through the power cable.

Power cable noise, especially alternator whine, can usually be reduced or eliminated by installing high-amperage filter chokes in *each* power lead (if the monster cables and a good RF ground at the transceiver haven't already knocked down the noise). Noise chokes are available from RadioShack and most car audio stores. I've sometimes had to use two noise filter chokes in series to eliminate persistent noise.

If the noise is being received through the antenna system, you might need to take more drastic measures to improve your RF ground. Hard-core operators, and those who insist on noiseless mobile environments, often install jumpers made from copper braid (the ends of each jumper are tinned), fastened with self-tapping sheet metal screws and star washers. Where do the jumpers go? Between every door, hood, trunk and body panel seam. Some even add jumpers between frame sections and between the frame and the exhaust system! Hey, it works!

If your antenna, mounted on the rear bumper, is receiving ignition noise, copper braid straps that connect the car body/frame to the hood panel (the jumpers are located under the hood and near the hinges) may eliminate the problem. It may defy logic, but it often works. By similarly bridging the doors, the trunk lid, and so on, you'll improve your antenna performance by providing a better RF ground, and you just might eliminate that last pesky noise problem to boot.

Installing copper braid jumpers can be a bit time consuming, but the results are worth the effort. Remember to "freshen" the jumper connections periodically. As soon as you install them they start to degrade. Don't wait until your system starts to get flaky!

Some pulsed ignition noise can be tamed by replacing your car's spark plugs with special low-noise "resistor plugs." Similarly, some spark plug cables can be replaced with low-noise variants.

Interference *to* your car can be merely annoying or potentially fatal. A road trip between central Minnesota and Fargo, North Dakota, highlighted this kind of problem for me. Every time I keyed my 2-meter hand-held, the cruise control would drop out, giving the car an unexpected jolt. This was more annoying than anything else. Some operators have zapped their car's on-board computers (big bucks) and fried the in-glass defroster wires in the rear windshield (which were "receiving" RF)! Treat these kinds of RFI situations with respect, call your dealer and consult *The ARRL RFI Book*.

VHF/UHF

VHF/UHF installations are pretty much like their HF counterparts—only easier. At these higher frequencies your car's body (or even a large panel) makes for a better RF ground. Magmount antennas often work without ground straps and additional gymnastics. Antennas are small, and some can even be mounted on your car's windows. If you fudge a bit on the antenna mount, be sure to follow proper power cabling procedures. Shorted power cables are dangerous no matter what!

TIPS AND TIDBITS

There's a lot to learn about effective mobile installations and operation. Here are a few points to ponder:

- An effective noise blanker is worth its weight in gold. When shopping for a mobile rig, test noise blankers carefully and check out *QST*'s Product Review section to see what the reviewers think about noise blanker performance. A good noise blanker can work wonders. A poor blanker will do *nothing*.
- Turn off your rig's receive preamp (and try adding 10 dB of attenuation). This will eliminate a lot of crud and weak signals that you're not likely to work anyway.
- Consider installing a good-quality auxiliary speaker. Your rig's speaker may sound great at home, but in a noisy mobile setting its tiny speaker may be muffled or aimed under the dashboard. If you crank up the gain distortion sets in. A speaker designed for mobile work, perhaps mounted near your head, can pump out crisp, clear, *loud* audio without breaking up.
- Don't wipe out your battery. A mobile rig that puts out 100 W can easily draw 25 A during transmit. Keep your engine running

while you're on the air or keep engine-off transmissions to a minimum. And be sure to turn your rig off when you exit the vehicle. Because it's connected directly, if you forget to turn it off it may drain your battery.
- Be safe, not sorry. Cell phones are handy, but drivers who use them are a known menace when it comes to car accidents. Enthusiastic ham operators can't be far behind, especially as we're tuning antennas, tweaking knobs, reading SWR meters, jotting logbook entries, and so on. Drive first, operate second! Keep your eyes and your mind on the road.
- Stop! Thief! Mobile rigs are juicy targets for thieves of all caliber. Keep your gear secure by locking your car or shuttling your gear to the trunk (for overnights or extended periods away from the vehicle). And make sure your insurance covers your gear *before* filing a claim. Most homeowner's and renter's policies *specifically exclude* electronic goodies that are in your car but can be powered by your car. Call your agent and get a rider that lists your gear. ARRL members should take advantage of the excellent ham-equipment insurance package available to them.

SUGGESTED READING

Burch, Roger, WF4N, *Your Mobile Companion*, ARRL Order no. 5129.

Burch, Roger, WF4N, "*You* Can Operate HF Mobile," *QST*, Feb 1993, p 29.

Ford, Steve, WB8IMY, "Going Mobile," *QST*, Dec 1991, p 23; Jan 1992, p 53.

Gold, Jeff, AC4HF, "HF Mobiling—Taking it to the Streets," *QST*, Dec 1993, p 67.

Hare, Ed, W1RFI, *The ARRL RFI Book* (Newington, CT: ARRL, Inc), ARRL Order no. 6834.

Johnson, Don, W6AAQ, *Everything You Forgot to Ask About HF Mobileering* (Sacramento, CA: Worldradio Books).

KA6WKE's HF Mobile Web Pages, **www.qsl.net/ka6wke/hfmobile.html**.

Seybold, John, KE4PRC, "HF Mobile Installation Tips," *QST*, Dec 1995.

WB6HQK's Mobile Antenna Shootout Results, **http://people.delphi.com/cecilmoore/shootout.htm**

Chapter 8

Radio Destinations

By now you know that as a Stealth operator you have a wide variety of operating options depending on your exact circumstances. We've spent a lot of time on setting up a home station, antennas and the basics of mobile hamming. What remains is something that I hinted at in Chapter 2—an introduction to *portable* hamming. Unlike many things, when it comes to ham radio, you *can* take it with you!

When most people conjure an image of portable operating, Field Day comes to mind. Actually, that's not a bad way to start. Field Day—thousands of operators working an emergency preparedness event from a variety of locations away from home—incorporates every aspect of portable hamming. But working Field Day is only the beginning.

With a little ingenuity, today's miniaturized electronics and the basics of Stealth Amateur Radio, you can take your ham radio hobby just about anywhere. Picnics, camping trips, road trips, a weekend at grandma's or a business trip to Tokyo (with an appropriate reciprocal ticket)—ham radio can go anywhere. And don't forget boat rides, hiking excursions and that summer you've always wanted to spend at the lake cabin.

SETTING UP SHOP

Deciding where to operate depends on where you are and what you're doing. Covering the range of possibilities is a bit broad for our treatment here. Just remember that the basics of Stealth Radio still apply. Unbalanced antennas still need RF grounds or counterpoises, higher antennas work better than lower antennas, and so on. The thing that's different is your location. Instead of being in your home shack you're out in the boonies somewhere.

There are, of course, a few general suggestions. Hilltops are pretty good for just about any radio activity, especially VHF/UHF. HF operators will want at least a few tall trees for stringing antennas, while VHF/UHF operators may have better luck if there are *only* a few trees (or even none) to absorb precious higher-frequency signals.

Remember to show the proper respect for the land (and the landowners) when you set up a portable station or campsite on property that's not your own. Don't break branches when stringing antennas, take down any antennas you put up and don't leave *any* garbage or debris behind when you leave. Get permission ahead of time, if necessary.

It's ironic that operating at a crowded campground can get you into the same RFI hot water that you soak in at home. Firing up on 20 meters may wipe out your fellow campers' radios and TVs. Be reasonable!

RIGS

Any rig will work from the field as long as you can supply the required power, but most hams who operate portable do so with compact commercial mobile rigs, commercial or kit QRP transceivers, or QRP gear they've built themselves.

Because most portable operation is done without access to the ac mains, most operators go for gear that runs on 12 V dc. Unless you're hiking or biking, you probably have a source of 12-V power nearby (car alternators, automotive or marine batteries, and so on). AC operation from battery power is possible with today's efficient "pure sine wave" inverters, which are more than 90% efficient in converting 12 V dc into smooth 120 V ac. Although high-quality inverters are somewhat expensive, avoid cheap inverters that put out sawtooth waves or "modified sine waves." They may work okay—or they may trash your expensive radios! See **Figure 8.1**.

When relying on battery power, transceiver characteristics that can be ignored at home can be quite important in the field. One of the most important is power consumption. If you're operating from a battery that can't be recharged until you get home (a common situation for portable operators), your rig's power draw will determine your operating time. Reducing your transmitter power can make a big difference when you're *transmitting*, but will do nothing to save power while you're receiving—which accounts for the bulk of your operating time.

To address this situation, some manufacturers have produced rigs that are designed for minimal current consumption while receiving. For example, most Wilderness Radio portable QRP rigs consume very little power in receive mode, and the Elecraft K2 is a complete 10 W HF CW/SSB transceiver with an optional built-in battery!

SGC's SG-2020—a low- to medium-power multi-mode, multiband HF transceiver designed from the ground up for field operations—treads very lightly in receive mode. And the '2020 has other benefits for portable operators. The radio is relatively

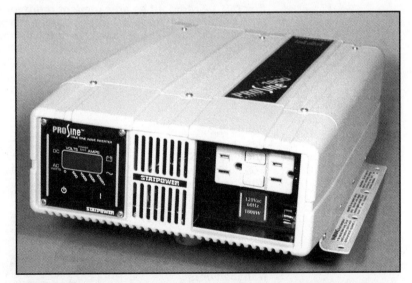

Figure 8.1—Although this 1800-W Statpower inverter is too large for most portable Stealth operations, its pure sine wave output is perfect for powering computers, transceivers and solid-state devices that may be damaged or destroyed by inverters that put out sawtooth waves or "modified sine waves." (author photo)

Wayne Burdick, N6KR, and Eric Swartz, WA6HHQ, designed the Elecraft K2 radio kit as an excellent radio for portable operation. With the optional built-in battery, antenna tuner and portable whip antenna, this all-band HF radio is ready to go anywhere. (WA6HHQ photo)

Figure 8.2—Designed for field operations, SGC's SG-2020 draws very little power while in receive mode—something your batteries will appreciate! This 1 to 20 W multimode, multiband HF transceiver is water-resistant and quite rugged. (author photo)

moistureproof and is built like a tank. You could probably run over the radio *with a tank* and not even dent the ruggedized milspec housing. See **Figure 8.2**.

ANTENNAS

Antennas for field use should be lightweight and unobtrusive. You don't want to mar your (or anyone else's) scenic vistas with a rat's nest of wires, feed lines or aluminum tubing. I've used two types of portable antennas over the years. One is a 40-meter dipole fed with 300-ohm TV twinlead. The elements are made from 20-gauge magnet wire, the center and end insulators are made from small, thin Plexiglas scraps, and the center and ends are held up with 30-pound-test monofilament fishing line. With a small tuner (with a built-in balun and an SWR meter) I can work all the bands from 40 meters and up.

My second portable antenna is even easier to set up and use. I simply toss a 66-foot wire into (or over) a tree and connect the near side right to the business end of my antenna tuner. I then roll out one or two 66-foot counterpoise wires and connect it (them) to my tuner's grounding post. This lazy vertical or inverted-L (depending on tree height, placement and density) *starts at the tuner*, which eliminates any loss from feed line runs. I can tune this antenna on all bands from 80 meters and up.

While using the vertical wire I haven't had any problems with "RF in the shack" at QRP levels, but it's occasionally troublesome at 50 W or so (not to mention potential RF exposure issues at power levels above 50 W). To make this antenna even easier to use I place an autocoupler at the base of the wire vertical and run a short coaxial feed line to my operating position. This is especially handy while operating from my mini camper. I have plenty of rechargeable 12-V power on hand to run the coupler, and I don't have to carry the tuner in a rucksack. The tuner I use is too big for backpacking, but smaller models—and those designed for portable/QRP use—are available in kit form from several manufacturers. **Figure 8.3** shows a close-up of the SGC SG-231 auto coupler.

Feel free to make a "portable" version of your favorite antenna. Remember to keep things simple, compact and lightweight. Portable

Figure 8.3—SGC's SG-231 autocoupler. The insulated post feeds RF from the internal tuning network while the post that runs through the metal backplane is the unit's RF ground. The antenna starts at these terminals—don't use even a short piece of coax between your autocoupler and your antenna's feed point! (author photo)

antennas don't have to last forever, and they don't have to survive hurricanes and winter storms, so don't be afraid to sacrifice ultimate survivability to achieve something that doesn't hog all of the space in your backpack!

FEED LINES

TV twinlead has always been a favorite for portable use. It has very low SWR losses, it's lightweight, it can be rolled into a small, flat package and it doesn't require special connectors. You'll need to use it with a tuner/balun, but you'll probably have that on hand anyway for multiband operation.

If you have the room and can stand the weight, conventional coax works in the field as it does at home. If you're thinking of using a mini coax such as the teeny RG-174, confine your efforts to 80 and 40 meters and keep the coax run as short as possible. Mini cables are just too lossy at higher frequencies or with long cable runs.

SAFETY

Before you put up any antenna—especially in unfamiliar areas—always double check for hidden power lines, telephone lines or other dangerous situations. If you see an overhead wire, always assume that it's hot. And be sure to follow the other common sense rules when setting up in the field. Don't place antennas where other people can come into contact with them. Route power cables and antenna leads so people can't trip over them. And keep your gear and any cables out of the water or away from wet areas.

YOUR PORTABLE INVENTORY

Determining what to take along on a portable radio excursion is a lot like packing for any other trip. The best possible preparation is to assemble the exact station you'll be using and put it on the air in your backyard before you leave town. Use the same antenna, the same battery, the same tuner, and so on. That way you'll know if you have everything you need when it's time to leave. When the station setup seems perfect, carefully make a checklist of your station's components and look it over while you pack items prior to departure.

A few additions to your barebones equipment list will accommodate an emergency or an unforeseen situation. As space and weight allow, consider bringing along a miniature logbook or notebook, a tiny digital multimeter, a pocketknife or multifunction Leatherman-style tool, electrical tape, extra wire, clipleads, a compact set of screwdrivers, a small wire cutter/stripper, a pair of Walkman-style headphones with an appropriate adapter—whatever you might need.

PORTABLE POWER

Providing power to our portable stations is an Achilles heel for many radio adventurers. Here's one area where a perpetual motion generator would come in very handy! If you're traveling by car or boat, you probably have a handy source of 12-V power along for the ride. But if you're hiking, biking or canoeing, for example, you'll have to carry batteries, a small generator or a bulky solar panel—none of which are all that appealing. When it comes to providing power there are no free lunches.

Basically, you have to scale your power requirements to match your available energy. For backpackers, hikers and those

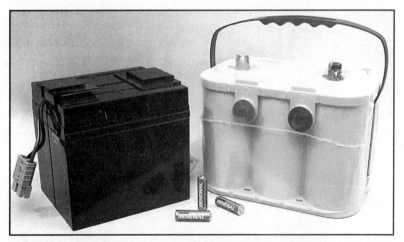

Figure 8.4—Shown here are examples of the most common batteries used for portable operation: A 12-V lead-acid, deep-cycle, marine/RV battery; a 12-V gel cell and 1.5-V nickel-cadmium (NiCd) rechargeables. Although the Optima deep cycle battery is designed for long life and deep discharges, it will be much happier staying at 50% charge or higher (as will any deep-cycle battery). (author photo)

"traveling light," a mini QRP rig designed for minimal (or micro) power consumption is a practical upper limit. Go beyond that and you'll exhaust your flashlight-size batteries in a jiffy.

If you can manage to carry a larger NiCd pack or a gel cell (and maybe a solar panel), a less exotic low-power rig will work just fine.

If you can handle a medium- to full-size deep cycle marine battery—recharged by a vehicle alternator, a compact gasoline-powered generator or a solar panel—the sky's the limit. That is, you can easily power your regular 100-W base station rig in the field. See **Figure 8.4**.

BATTERIES

Not too many years ago there were only a few types of rechargeable batteries. Now, however, thanks to steady increases in battery technology provided by periodic space races and energy crises, hams have an overwhelming array of choices when it comes to portable power. Entire books on rechargeable batteries are available. I can cover only a few basics here. Lithium-ion and nickel-metal hydride batteries, popularized by laptop computers, pack a lot of power into a small space, but they're very expensive and require specialized chargers. They won't be discussed here. See the 2000 edition of *The ARRL Handbook* for more information and additional resources.

Ni-Cds

Usually known as NiCds (pronounced "nye-cads"), nickel-cadmium batteries are usually packaged in small bricks or common consumer sizes (AAA, AA, C and D-size cells). Lumped in with NiCds are several new types of rechargeable consumer-grade alkaline cells. These batteries produce 1.2 to 1.5 V per fully charged cell. Connect them in series or parallel to get the voltage and current capacity you need to power your rig.

Without getting too far afield, battery capacity is measured in ampere hours (Ah). A 3-Ah battery can *theoretically* produce 3 A of current for one hour (or 1 A for three hours, a half an amp for six hours, and so on). In practice these figures are degraded by age, temperature, internal resistance, charge-discharge history, and other factors.

The amp-hour ratings of common NiCd cells range from about 0.2 (AAA cells), to 1.75 (D cells), to 5 or more (large camcorder bricks). As with all batteries, proper charging is a must. Typical NiCds are charged at a current equal to a tenth of their capacity in Ah. Be sure to follow the manufacturer's charging recommendations!

Lead-Acid Automotive and Marine Batteries

Twelve-volt car batteries—the often-abused underdogs of portable power—are designed to pump out several hundred amps in short bursts. They're poor choices for powering radio gear, which requires small, steady current draws over time. They're also extremely heavy. Garden-tractor batteries weigh 15 pounds or more, while large car batteries tip the scales at about 60 pounds. Car batteries are designed to accommodate large charging currents and can easily be destroyed if they're deeply discharged.

Twelve-volt marine batteries, on the other hand, are designed to power trolling motors and survive much deeper discharges. Deep-cycle batteries, as they're commonly called, are much better choices for portable radio power systems and those fed via generators and solar panels. Contrary to popular belief, the life of a deep-cycle battery can be greatly extended by never discharging it below 50% capacity. Typical marine batteries have capacities of 60 to 90 Ah and can be charged at 1 to 20 A, depending on size and depth of discharge.

Gel Cells

Gel cells are essentially space-age lead-acid batteries with a twist: They're completely sealed, so they don't leak, even if you tip them over. A gelled electrolyte replaces the liquid found in car and marine batteries. Gel cells, although more expensive, come in sizes from 2 to 50 Ah. They're more sensitive to charging conditions, so be sure to get a charger or a solar charge controller that's designed to handle them.

Solar Panels

Because of improved yields and manufacturing techniques, solar panels—collections of photovoltaic cells—are on the verge of cost effectiveness in certain residential and commercial applications where the ac mains are too far away or too costly to connect. Most commercial panels produce about 18 V dc with no load. The Solarex panels on my camper each produce 18 V at about 3.5 A. Although a large deep-cycle battery could handle that full-time, that's way too much charging power for most small batteries. (I use three panels to provide about 150 W of peak charging power. The panels could eke out a bit more juice, but my limited mounting arrangement doesn't track the sun.) See **Figure 8.5**.

Some hams use three-terminal regu-lators and big blocking diodes to prevent the battery from discharging through the panel at night. I recommend building or buying an inexpensive charge controller to tame the panel's output and take better care of your deep-cycle battery. See **Figure 8.6**.

Purchased new, 65 W (3 to 4 A) of peak solar charging power costs $300 to $400 (which includes the price of an entry level charge controller). Surplus panels are about half that, but their output power and efficiency may be degraded somewhat. Sixty-five watts of peak solar charging power weighs about 20 pounds and measures about 20 by 36 inches. Because solar panels are made of glass, you can't just bang them around, either. Solar power is great if you can accommodate it, but it's a pain if you can't.

Charging Batteries

Some operators charge car, marine

Figure 8.5—This Solarex panel puts out 18 V dc (at no load) and about 65 W of battery charging power in full sun. Although the efficiency of solar panels has improved greatly over the past decade, high-power panels are still a bit cumbersome. (author photo)

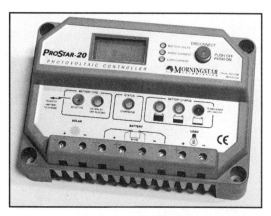

Figure 8.6—Designed to keep your batteries optimally charged—and to prevent them from discharging through your solar panel at night—charge controllers such as this Morningstar unit are recommended over simple blocking diodes, especially if you value your battery! (author photo)

and gel-cell batteries on any old charger they happened to buy at the local auto parts store. This can be a big mistake. As I learned when researching solar power systems for boats and RVs, properly charging batteries is a complicated process.

Depending on their exact electrolytes and internal designs, batteries require specific charging voltages. They also accept charging current in three distinct phases depending on how deeply they're discharged. To make matters worse, all of these variables change with temperature!

Garden-variety chargers spew out unregulated half-wave dc and let the battery draw what it will for charging current—and damn the temperature! Quality chargers provide regulated dc at the proper voltages and adjust their voltage and current outputs according to the state of the battery *as it charges*. The best units feature temperature sensors that keep the battery from over-heating and experiencing "thermal runaway" during the charge cycle. See **Figure 8.7**.

Unfortunately, "smart" chargers are more expensive than their "dumb" counter-parts (sometimes a lot more expensive). If you're using expensive batteries and you want to get a few years' use out of them instead of a single season, get a good charger. Many quality chargers, such as the RF-friendly, lightweight switching models made by Newmar and Statpower, are so clean they can double as high-current 12-V power supplies. The double duty can ease the pain of the purchase!

PORTABLE GENERATORS

Generators are good news-bad news items. Portable generators are shrinking—and that's good news for Stealth operators. Even if your favorite Field Day site is powered by a 5000-W monster generator, you don't want to drag one of those around even if you have a pickup at your disposal.

The bad news is that the ac output from most mini generators is poor harmonically and not sufficiently regulated to risk powering your expensive solid-state radio or computer gear directly. Some of the smallest units, however, such as those made by

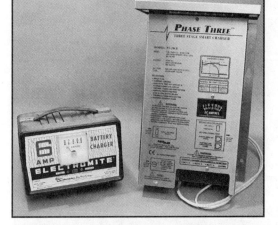

Figure 8.7—Battery chargers, old and new. If you're using expensive deep-cycle batteries and want them to last for more than a season or two, don't charge them with **any old charger. Modern chargers such as Newmar's PT-20 feature three-stage regulated charging, temperature compensation and remote control. Use a modern charger—your batteries will thank you!** (author photo)

Coleman, Honda, Kawasaki and Yamaha, among others, provide high-amperage 12-V battery charging outputs and weigh less than 30 pounds. **Figure 8.8** shows a homemade dc generator.

Unless you have specific knowledge or experience otherwise, use the dc generator outputs only for charging batteries—don't power your radio gear directly. The dc output is almost certainly unregulated, and it may be loaded with ripple, noise and other crud. You may be willing to sacrifice a cheap deep-cycle battery from your local auto parts superstore by charging it from such a generator. I would be reluctant to risk damaging an expensive battery this way, though.

In the right situations, mini generators can be an excellent source of "charged batteries." They do require gasoline, however, and even the relatively quiet ones can be *loud*. Keep these factors in mind before you fire yours up in a crowded campground!

MOUNTAINTOPPING

So, you say you can't afford a 3000-foot tower to support your VHF/UHF antennas and your hot air balloon is in the shop? Well, the next best thing is a nearby mountaintop—or at least a big hill (for readers in the flatlands). VHFers have been operating from mountaintops for years, often in conjunction with spring and

Figure 8.8—Commercial power generators usually produce 120-V ac or unregulated half-wave dc (for charging car batteries). Units suitable for powering solid-state devices tend to be bulky and heavy. If you've been searching for clean, portable dc power, why not build your own generator? Yaniko Palis, VE2NYP, did just that. Read about how to build his "12-V Pup" in June 1997 QST. (James St Laurent photo)

fall contest weekends and Field Day. You don't have to limit yourself to contests to enjoy mountaintopping, however. If your increased elevation gives you line of sight propagation to a distant metro area, you'll make plenty of contacts even if the bands aren't open for long-distance QSOs. See **Figures 8.9** and **8.10**.

Setting up a nifty stack of VHF/UHF antennas isn't too difficult because the antennas tend to be rather small compared to their HF cousins. A weighted roof tripod and a 10-foot mast can support several small beams. Some operators support their antenna masts by bracketing them onto the sides of their campers or by welding an inexpensive fixture—a flat metal plate with a two-foot piece of pipe welded near one edge. The pipe diameter is chosen so that the antenna mast just fits inside. With the plate positioned appropriately on the ground, users "drive over" the plate and park their vehicles so a tire remains on top. This provides a quick, inexpensive, sturdy and portable mast mount.

If you're using beam antennas, make sure your antennas can be rotated by hand or with a small TV rotator. Beam antennas have narrow beamwidths, so the ability to aim them can be critical. If you're not familiar with the area you might even want a map or two to figure out where to point your antennas.

In your rush to get on the air and work stations far and wide, don't forget the basics of RF safety. Make sure you and anyone nearby can't walk in front of your transmit-ting antennas. Running

Figure 8.9—Roger Hayward, KA7EXM, of Corvallis, Oregon, combines backpacking, contesting and QRP as he operates in the September 1991 ARRL VHF QSO Party with a 200-mW hand-held SSB/CW transceiver from the top of Oregon's South Sister (10,385 feet). In June 1992 QST, Roger describes his other "road trip" radio—a 1.5-W 40-meter CW rig. (photo by Scott Brown, WB7SHE)

Radio Destinations 8-13

Figure 8.10—Ah, the good life—contesting in the Colorado Rockies! Looking north from North Cochetopa Pass, Bob Witte, KBØCY, searches for contacts in the ARRL June VHF QSO Party. In contest lingo Bob is a rover. His Jeep (and the small size of his VHF/UHF antennas) allows him to quickly set up, operate, take down and sprint to another remote location.

even low power to high-gain antennas produces lots of effective radiated power.

Some contest operators build VHF/UHF "superstations" into four-wheel-drive vehicles so they can quickly move between remote mountaintops. These hams, known as "rovers" in VHF contests, put rare, uninhabited grid squares on the map and are highly sought-after during contests. Building ever-more-capable rover stations can be habit forming. You were warned!

In a more relaxing light, mountain-topping can be whatever you make it. If you don't want to take a bunch of gear to a parking lot or rest area that's perched next to a scenic overlook, take your hand-held transceiver along on your next hike into the hill country. As you crest each new hilltop you're mountaintopping!

SUGGESTED READING

Andera, Jim, WBØKRX, "8-Band Backpacker Special," *QST*, Jun 1994, p 68.

Brook, Dick, KJ1O, "Two-Meter Mountaintopping…Sort Of!" *QST*, Feb 1993, p 58.

Casier, David, KEØOG, "Solar Power for Your Ham Station—It's Easier than You Think," *QST*, Apr 1996.

Danzer, Paul, N1II, Ed., *The ARRL Operating Manual* (Newington, CT: ARRL), 6th Edition, 1997.

Kennedy, Dennis, N8GGI, "A Packable Antenna for 80 through 2 Meters," *QST*, Apr 1997.

Kleinschmidt, Kirk, NTØZ, "How to Choose and Use a Portable Power Generator," *QST*, Jun 1999.

Nerger, Paul, KF9EY, "Roving for Gold in the Colorado Rockies," *QST*, Jun 1993, p 21.

Palis, Yaniko, VE2NYP, "The 12-Volt Pup: A DC Generator You Can Build," *QST*, Jun 1997.

Putman, Peter, KT2B, "Go Tell it on the Mountain," *QST*, May 1990, p 15.

Putman, Peter, KT2B, "Mountaintopping VHF Operating: A New Adventure," *QST*, Oct 1993, p 60.

About the ARRL

The seed for Amateur Radio was planted in the 1890s, when Guglielmo Marconi began his experiments in wireless telegraphy. Soon he was joined by dozens, then hundreds, of others who were enthusiastic about sending and receiving messages through the air—some with a commercial interest, but others solely out of a love for this new communications medium. The United States government began licensing Amateur Radio operators in 1912.

By 1914, there were thousands of Amateur Radio operators—hams—in the United States. Hiram Percy Maxim, a leading Hartford, Connecticut, inventor and industrialist saw the need for an organization to band together this fledgling group of radio experimenters. In May 1914 he founded the American Radio Relay League (ARRL) to meet that need.

Today ARRL, with approximately 170,000 members, is the largest organization of radio amateurs in the United States. The ARRL is a not-for-profit organization that:
- promotes interest in Amateur Radio communications and experimentation
- represents US radio amateurs in legislative matters, and
- maintains fraternalism and a high standard of conduct among Amateur Radio operators.

At ARRL headquarters in the Hartford suburb of Newington, the staff helps serve the needs of members. ARRL is also International Secretariat for the International Amateur Radio Union, which is made up of similar societies in 150 countries around the world.

ARRL publishes the monthly journal *QST*, as well as newsletters and many publications covering all aspects of Amateur Radio. Its headquarters station, W1AW, transmits bulletins of interest to radio amateurs and Morse code practice sessions. The ARRL also coordinates an extensive field organization, which includes volunteers who provide technical information for radio amateurs and public-service activities. In addition, ARRL represents US amateurs with the Federal Communications Commission and other government agencies in the US and abroad.

Membership in ARRL means much more than receiving *QST* each month. In addition to the services already described, ARRL offers membership services on a personal level, such as the ARRL Volunteer Examiner Coordinator Program and a QSL bureau.

Full ARRL membership (available only to licensed radio amateurs) gives you a voice in how the affairs of the organization are governed. ARRL policy is set by a Board of Directors (one from each of 15 Divisions). Each year, one-third of the ARRL Board of Directors stands for election by the full members they represent. The day-to-day operation of ARRL HQ is managed by an Executive Vice President and his staff.

No matter what aspect of Amateur Radio attracts you, ARRL membership is relevant and important. There would be no Amateur Radio as we know it today were it not for the ARRL. We would be happy to welcome you as a member! (An Amateur Radio license is not required for Associate Membership.) For more information about ARRL and answers to any questions you may have about Amateur Radio, write or call:

ARRL—The national association for Amateur Radio
225 Main Street
Newington CT 06111-1494
Voice: 860-594-0200
Fax: 860-594-0259
E-mail: **hq@arrl.org**
Internet: **www.arrl.org/**

Prospective new amateurs call (toll-free):
800-32-NEW HAM (800-326-3942)
You can also contact us via e-mail at **newham@arrl.org**
or check out *ARRLWeb* at **http://www.arrl.org/**

About the Author

First licensed in 1977 at age 14 (as WDØBDA) I spent most of my early years chasing states and DXCC countries on CW. My parents, grateful that I was passionate about a "respectable" hobby, tolerated my maze of backyard antennas and my sprawling "mad scientist" basement shack.

My freshman year in college coincided with a solar cycle peak and marked the beginning of my career in Stealth Amateur Radio. Most of the stations and antennas I used during those "Stealth Years" are chronicled in this book. A few others will remain forever unmentioned!

In May of 1988, I leveraged my experience in broadcast journalism and my years as a ham operator and joined the ARRL HQ staff as an assistant technical editor. After working on the 1990 *ARRL Handbook*, I moved to the other end of the building and accepted a position as *QST*'s assistant managing editor (and *de facto* staff photographer).

I have many fond memories of my six years at ARRL HQ, but the woods and water of Minnesota were calling me home. In July of 1994 I moved back to my hometown of Litle Falls (only 30 miles north of Lake Wobegon). In addition to my periodic work for *QST*, I write a ham radio column for *Popular Communications*, I syndicate features on small office technology to more than a hundred newspapers nationwide, I do a fair amount of freelance technical writing for area industries, and I dabble in computers.

Although I again live in an area that is virtually free of covenants and deed restrictions—and I don't need to worry about Stealth antennas—you'd be hard pressed to find the large horizontal loop I've strung between the tops of three tall trees (the 10-foot satellite dish in the backyard is much more obvious). Because QRP has been my thing for years, maximum smoke for me is 100 W. With my big loop I can work almost everything I can hear from 160 through 6 meters, so I have no compelling need for "better" antennas.

Based on my experience, a Stealth—or semi Stealth—approach is a great way to enjoy ham radio. I hope you'll give it—and ladder line feeders—a try.

Good luck, and 73,

Kirk Kleinschmidt, NTØZ

FEEDBACK

Please use this form to give us your comments on this book and what you'd like to see in future editions, or e-mail us at **pubsfdbk@arrl.org** (publications feedback). If you use e-mail, please include your name, call, e-mail address and the book title, edition and printing in the body of your message. Also indicate whether or not you are an ARRL member.

Where did you purchase this book?
☐ From ARRL directly   ☐ From an ARRL dealer

Is there a dealer who carries ARRL publications within:
☐ 5 miles   ☐ 15 miles   ☐ 30 miles   of your location?   ☐ Not sure

License class:
☐ Novice   ☐ Technician   ☐ Technician Plus   ☐ General   ☐ Advanced   ☐ Amateur Extra

---

Name _____   ARRL member? ☐ Yes ☐ No
                                                         Call Sign _____
Address _____
City, State/Province, ZIP/Postal Code _____
Daytime Phone ( ) _____   Age _____
If licensed, how long? _____   e-mail _____
Other hobbies _____
_____
Occupation _____

For ARRL use only	STEALTH
Edition	1 2 3 4 5 6 7 8 9 10 11 12
Printing	3 4 5 6 7 8 9 10 11 12

From _____

_____

_____

| Please affix postage. Post Office will not deliver without postage. |

EDITOR, STEALTH AMATEUR RADIO
ARRL—THE NATIONAL ASSOCIATION FOR AMATEUR RADIO
225 MAIN STREET
NEWINGTON CT 06111-1494

— — — — — — — — — — —   please fold and tape   — — — — — — — — — — — —